I0794465

IMAGES
of America

SOUTHEASTERN ARIZONA MINING TOWNS

Perhaps no better representation of mining interests in the history of southeastern Arizona can be found than Tucson's official coat of arms. Designed in 1929 by Capt. Donald W. Page, the city building inspector, with the help of fire chief Joseph A. Roberts, Fr. Victor C. Stoner, and Mexican artist Jesus V. Zubiade, the coat of arms represents the four major historic political divisions in Tucson, including the Spanish Conquest, symbolized by the Pillars of Hercules; the Mexican eagle depicting the period of Mexican domination; the American eagle representing the sovereignty of the United States; and finally Arizona statehood symbolized above with a radiant star. Minor epochs represented in Tucson's history include the Jesuits and Franciscans with their arms of the Society of Jesus and the Order of St. Francis of Assisi. The presidio period and Apache warfare in the region appear with a wall overtopped by a black hill, while the battle flag of the Confederacy represents Tucson's role as the confederate capital in Arizona territory. All countries, participants, and groups represented in Tucson's coat of arms contributed to mining the mineral wealth in southeastern Arizona. The Tucson City Council adopted the coat of arms on February 19, 1929. (Photograph by the author; courtesy of the Arizona Historical Society.)

ON THE COVER: Surface equipment belonging to the Inspiration Consolidated Copper Company is evident in this image. Specifically, a hoist, compressor house, and coarse crushing plant positioned at the main shafts can be seen. This photograph was taken looking south toward the Pinaleno Mountain Range in 1911. By 1915, the first large-scale application of flotation was first used at the Inspiration Mine in Miami, Arizona. (Courtesy of US Geological Survey.)

IMAGES
of America

SOUTHEASTERN ARIZONA MINING TOWNS

William Ascarza

ARCADIA
PUBLISHING

ISBN 978-0-7385-8516-1

Published by Arcadia Publishing
Charleston, South Carolina

Printed in the United States of America

Library of Congress Control Number: 2011936454

For all general information, please contact Arcadia Publishing:
Telephone 843-853-2070
Fax 843-853-0044
E-mail sales@arcadiapublishing.com
For customer service and orders:
Toll-Free 1-888-313-2665

Visit us on the Internet at www.arcadiapublishing.com

To my lovely wife, Lorie, who has been instrumental in supporting my endeavors to complete this book.

Contents

ACKNOWLEDGMENTS

I would like to thank the following individuals and institutions for providing me with access to historical information and print usage: Albertype Company; American Institute of Mining Engineers; Arizona Board of Regents for the University of Arizona; Arizona Historical Society; Arizona-Sonora Desert Museum; Arizona Geological Survey; Marilyn and Emmanuel Ascarza; Mark Ascher; Stacia Bannerman; Kevin Bays; Stan and Connie Benjamin; Marta Berry; Diane Braswell; Wendall Brown; Tallia Pfrimmer Cahoon; Mark Candee; Dan Caudle; Charles Cavalli; Lorie Cavalli; Kenny Don; Jez Gaddoura; Gleeson, Arizona, Museum; Greenlee County Historical Society; Jared Jackson; Peggy Larson; Dan Lee; Library of Congress Prints and Photographs Division; Alvin F. Lindstrom; Mike Litchfield; Marylou Long; Magma Copper Company; Pat McClain; Tina Miller; Katherine Reeve; Hector Rodriguez; San Pedro Valley Arts & Historical Society; Gene Schlepp; Cosme Rodriguez Sesma; Ken Shake; Patty Sirls; Glenn Snow; Southern Methodist University, Central University Libraries, DeGolyer Library; Town of Mammoth; University of Arizona Mineral Museum at Flandrau Science Center; Tucson Gem & Mineral Society; US Geological Survey (USGS); US National Archives and Records Administration; and Richard Langley Ware Their generosity and time was invaluable in making this book a realisty.

INTRODUCTION

Arizona has a rich mining history dating back over 1,000 years. The indigenous people known as the Hohokam, or "vanished ones," were the first to exploit the vast mineral resources in the landmass known today as Arizona. They used minerals such as copper and turquoise for ornamental jewelry and to trade among settlements. There is evidence that the Tohono O'odham (Papago) Indians mined hematite for use as war paint in the 15th century, shortly after the disappearance of the Hohokam. Native Americans were involved in mining turquoise in the Cerbat Mountains and cinnabar in the Castle Dome Mining District near Yuma. They also mined salt near Camp Verde. Although they were the first to mine the surface of Arizona, it was the Spanish who were the first to extensively penetrate its earth in search of mineral wealth, most notably in southeastern Arizona.

The Spanish first entered the region that was later called Arizona in the early 16th century. Their mission was to obtain the "Three Gs"—glory, God, and gold–for the Spanish crown. Early Spanish exploration of Arizona began with the expedition led by Fray Marcos de Niza in 1539. The following year, his reports of great wealth in the form of gold and silver reached Francisco Vasquez de Coronado, who mounted several expeditions aimed at discovering the "Seven Cities of Gold," which were rumored to rival the Aztec and Inca gold caches in Mexico and South America. The mineral wealth of the fabled "Seven Cities of Gold" proved elusive for the Spanish; however, they did succeed in establishing distant mining claims across the Southwest, including in southeastern Arizona.

During the time of Spanish rule, advancements were made in mining and refining minerals. The Spanish used an *arrastra* to pulverize ore deposits of gold and silver. Run by animal power, usually a horse or mule, the raw ore was crushed, sometimes amalgamated, and later sent to a sluice to separate the mineral content. The Spanish introduced the process of amalgamation to the New World in the 16th century as a means of separating silver from its ore. Mercury is mixed with silver removing it from its ore, forming amalgam. The applications of heat or nitric acid remove the mercury from the silver. Placer mining, which used techniques such as panning and the sluice box, was also conducted by the Spanish as a low-cost alternative for finding mineral wealth.

The arrival of the Jesuit missionary Eusebio Francisco Kino in the 1690s, along with the added protection of recently established missions of Guevavi, San Xavier del Bac, and Tumacacori, gave the Spanish impetus to explore the region for precious metals. According to the accounts of Father Kino, the Tubac-Tumacacori area was mined by both the Spanish and the Papago Indians for rich veins of gold and silver. Some sources speculate that the origin of the name *Arizona* may have been derived from a mining district called Real de Arizonac, which was located in northern Sonora and southwest of the present town of Nogales. The mine produced an abundance of silver during its operation in the late 1730s. Some silver specimens weighed in excess of 2,500 pounds. The area later became known as Planchas de Plata ("slabs of silver") or Bolas de Plata ("balls of silver").

Mining proved a risky venture in many parts of Arizona until the late 19th century because of the danger posed by the Apaches and Navajos. The Spanish continued to mine the region during a 30-year reprieve of hostilities brought about by Viceroy Bernardo de Galvez's introduction of the "Peace by Deceit" plan. In return for a cessation of hostilities against the Spanish, the Native Americans were given food, guns, and whiskey. Hostilities in the region resumed after Father Miguel Hidalgo's "Grito de Delores," which initiated Mexico's long war for independence from Spain. The Mexican Revolution of 1822 further reduced military protection for miners in the American Southwest against the onslaught of deprivation perpetrated by Apaches and outlaws.

Manifest Destiny brought American mining interests into Arizona in 1847. Tom Childs Jr. led a party of 19 Americans into the region on the quest to discover Planchas de Plata in search of the fabled cache of silver. He discovered abandoned copper mines in Ajo before his expedition was evicted by Mexican inhabitants. However, the discovery of gold in California brought about a resurgence of mining activity in the American Southwest. The combination of land acquired from Mexico after the Mexican-American War and the Gadsden Purchase of 1853 brought the remaining land comprising southeastern Arizona under the jurisdiction of the United States. The Mining Law of 1872, enacted by Congressional authority, increased mining activity in southeastern Arizona by granting US citizens the opportunity to acquire land for mining purposes for a small fee.

The Americans drew upon many sources of mining to develop Arizona's mineral deposits. Beginning with the site of Arizona's first gold rush in the defunct town of Gila City, mining became more prevalent across the state with additional major gold strikes found in La Paz, the Bradshaw Mountains, Rich Hill, Vulture Mine, and Greaterville. After the War Between the States, the value of silver increased, and so too did the mines in Arizona. Some of the more prominent silver mines in Arizona during the latter half of the 19th century included the Silver King Mine in the Pinal Mountains and Tombstone, both found in southeastern Arizona.

The Spanish and early Americans were initially drawn to the region by gold and silver, but copper mining soon became attractive, especially in the southern portion of the state. Demand for copper increased by the beginning of the Industrial Revolution in the United States, because copper is a good conductor of electricity and has many uses, including pipes, electrical wiring, machinery, transportation, and coinage.

By the 20th century, mining towns in southeastern Arizona, such as Bisbee, Clifton, and Ray, were well established and noteworthy for their large deposits of copper. Technology revolutionized the mining industry with the advent of pneumatic drills, which replaced the standard single- and double-jacking techniques employed by miners during previous centuries. Block caving, which is a mass mining method for extracting large bodies of lower-grade ore, and open-pit mining revolutionized the mining industry. Advances in the refinement of mineral extraction led to the ability to remove greater amounts of copper from otherwise lean ore deposits. The flotation process, cyanide leaching, and electrowinning are examples of some 20th-century advancements in copper production.

The 1906 strike in Cananea, Mexico, certainly had its influence in the region. Many of the participants on the American side were from southeastern Arizona and had a vested interest in the mines of Cananea. The Arizona Rangers were called in by Col. William Greene, the president of the Cananea operation and southeastern Arizona miner, cattle rancher, and land mogul, to quell the strike. Although the strike was forcibly ended, the repercussions would be far reaching, leading to the Mexican Revolution several years later and serving as the impetus for future labor strikes in Northern Mexico and southeastern Arizona.

Despite many decades of tension throughout the 20th century, relations between miners and mining companies have improved over the past 100 years. During the 18th and 19th centuries, miners had to work under primitive conditions for 10 to 12 hours a day, six days a week. Wages were lucrative, although most were spent by the miners at the company store or the gambling hall to buy spirits and food and partake in ventures with the ladies of the night. By the early 20th century, labor unions began to gain a foothold in the state, much to the disapproval of mine managers. During the period of 1884 through 1917, conflict between miners affiliated with the union and mine companies resulted in strikes, firings, and deportation. However, improvements were made, such as a reduction in daily work hours, medical benefits, and salaried positions, adding to the more unified workforce present in Arizona mining operations today.

Today, copper mining remains a major industry in southeastern Arizona and is one of the state's renowned "Five Cs," which include cattle, citrus, climate, and cotton. Over 400,000 documented mining claims have existed in Arizona since the arrival of the Europeans in the 16th century. A total of 60 percent of the copper produced in the United States owes its provenance to Arizona. Active, inactive, and future mining localities will continue to enhance the economy and history of southeastern Arizona.

One

Beginning Prospects

Beginning with the early prospects of Native Americans through the Industrial Revolution in the United States, mining in the geographical area known today as southeastern Arizona became an indispensable means for establishing the local economy, whether among tribes, mining camps, or towns. Native Americans known as the Hohokam operated mines around 1,000 AD in search of clay, coal, turquoise, and salt. They established trade routes between distant tribes for the purpose of obtaining commodities of value that were scarcer in their home region.

By the 16th century, several hundred years after the disappearance of the Hohokam, the Spanish actively mined the area, discovering large silver deposits for their king while establishing colonies on their ultimate journey to found San Francisco. Spanish explorers, such as Coronado, Father Kino, Capt. Juan Bautista de Anza II, and Fray Marcos de Niza, believed that the land today known as southeastern Arizona contained abundant mineral wealth. During the mid-19th and early 20th centuries, the United States followed up on the successes of the Spanish, developing previously mined prospects and discovering new deposits of gold, silver, and, most importantly, copper. Mining towns in southeastern Arizona significantly contributed to Arizona's economic development and eventual statehood on February 14, 1912.

Railroads played an integral part in the mining history of southeastern Arizona, connecting both larger and smaller mining towns to the mill, the smelter, and the product's final destination, to market. By 1880, the Southern Pacific Railroad arrived in southeastern Arizona, establishing the towns of Willcox and Benson while connecting Tucson to the outside world. Over the next century, the railroad continued to deliver people and supplies to the mining towns of Southeastern Arizona.

While the railroad connected the mining towns to the outside world, the towns themselves developed schools, general stores, boardinghouses, jails, recreation halls, and a strong sense of community. Paper script and tokens were freely exchanged among the townspeople and the company store run by the mine. Today, many of these early mining towns have long since been abandoned and have been dismantled, vandalized, or eroded through the passage of time. However, their story remains to be told through documented literature and photography.

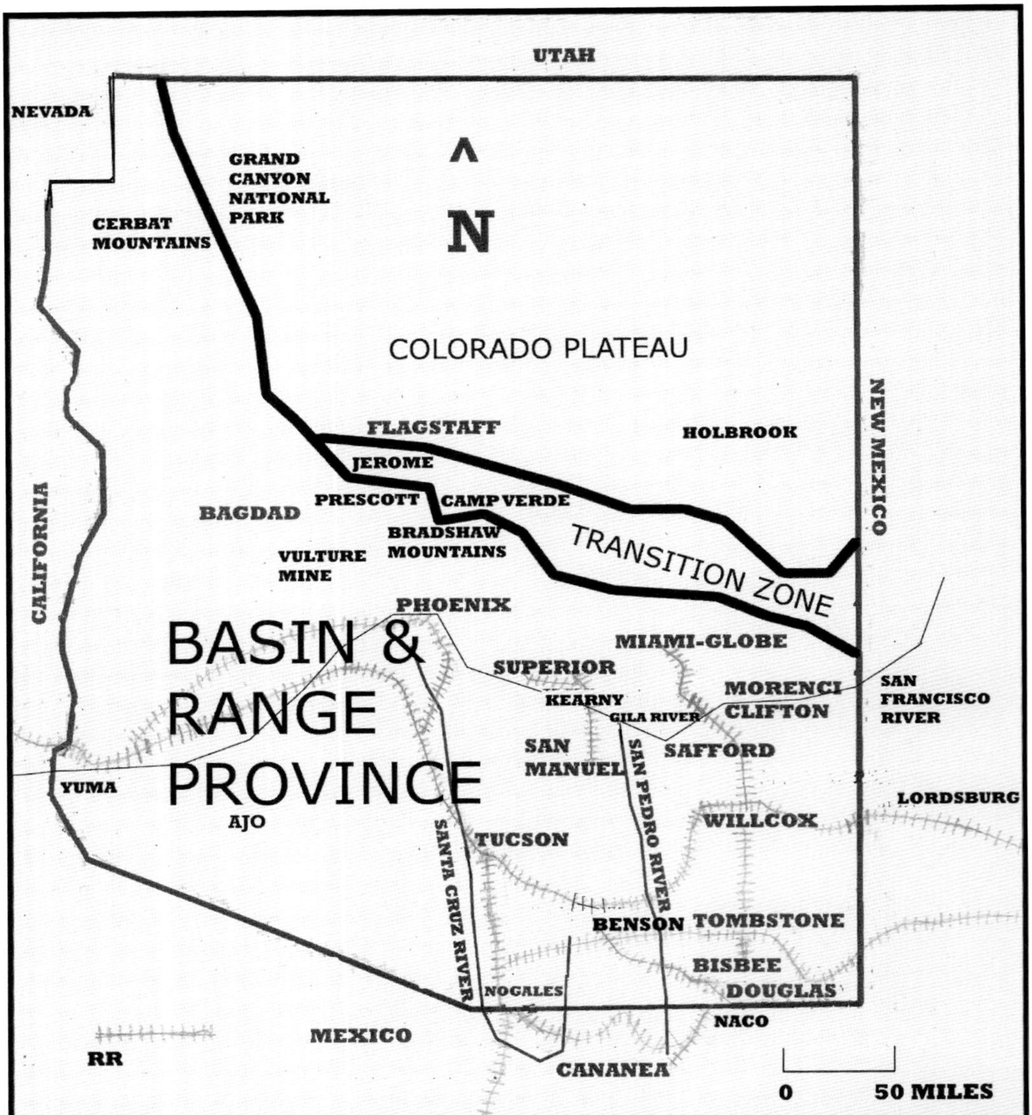

This map depicts Arizona and the major mining operations in its southeastern region. Included on the map are major railroads used to transport ore to the mill, smelter, and to market. Some railroads include the Arizona Eastern Railroad, Tucson & Nogales Railroad, Clifton & Southern Pacific Railroad, Tucson, Globe & Northern Railroad, and the Magma Arizona Railroad. The arrival of the railroad enhanced mining interests throughout Arizona connecting it to the lucrative financial markets both in the East and in California. (Courtesy of the author.)

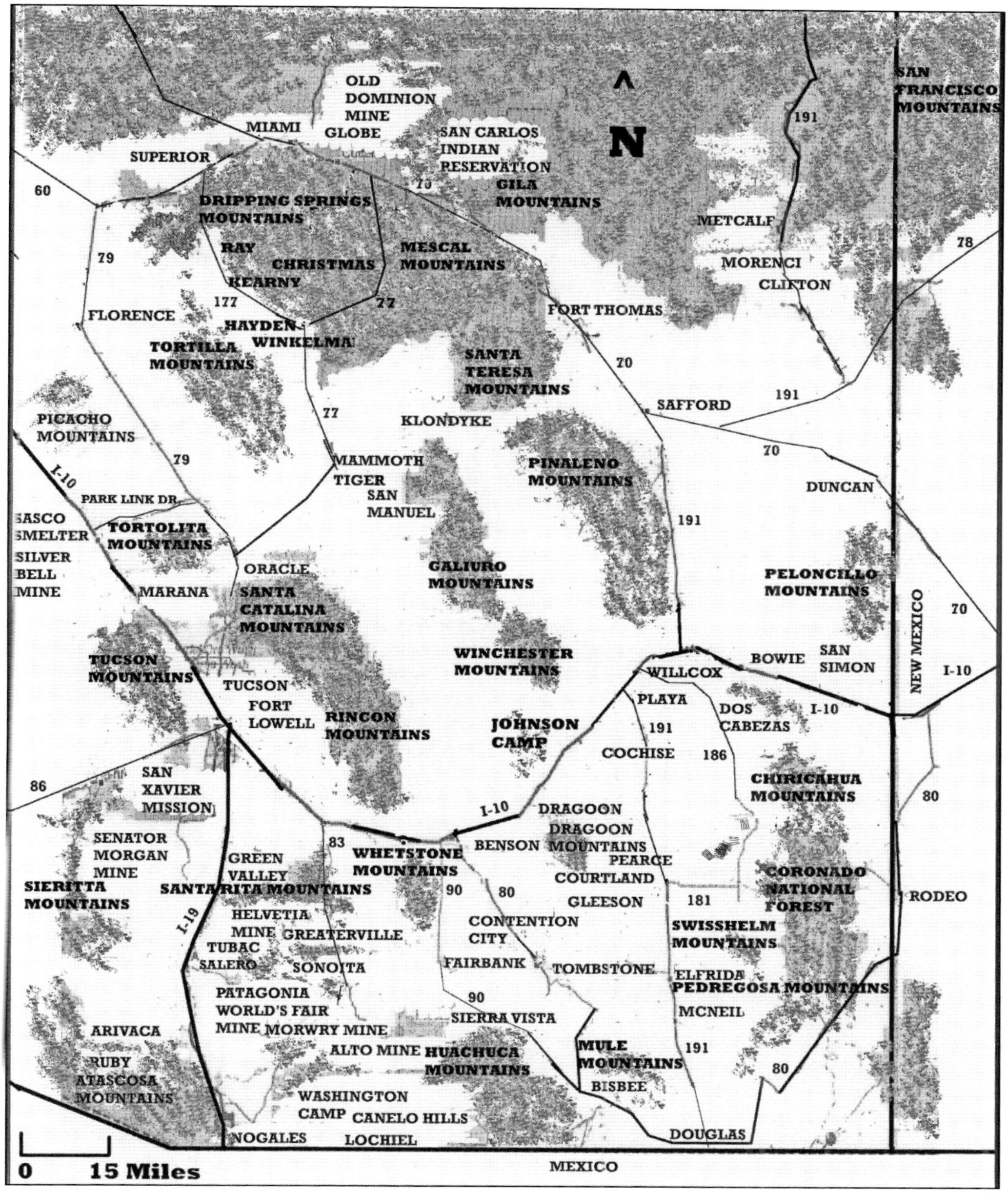

This map depicts minor mining localities in southeastern Arizona. It also includes the highways and interstate system as they appear today. The unique Basin and Range geologic features are also evident in this map consisting of a series of mountain ranges divided between valleys. (Courtesy of the author.)

Capt. Francisco Vasquez de Coronado traversed along the San Pedro River in 1540. His failure to discover the Seven Cities of Gold, coupled with his inability to reach a waterway through North America, forced him to return to Mexico City two years later and doomed his mission in the New World. Coronado's name lives on with both Coronado National Forest and Coronado National Monument. (Courtesy of Library of Congress Prints and Photographs Division.)

This monument, located near the Mexican-American border in the town of Lochiel, Arizona, pays homage to Fray Marcos de Niza. De Niza was a European explorer who entered Arizona on April 12, 1539, the first European west of the Rockies. He also provided a false report of a gleaming city of gold called Cibola to Viceroy Antonio de Mendoza that would ultimately lead to the ill-fated Coronado expedition. (Courtesy of the author.)

Father Kino, seen here with an Indian guide, is on the move. Traveling with his cattle, horses, and soldiers to multiple missions in southeastern Arizona, Padre Eusebio Francisco Kino, an Italian-born Jesuit, was a renowned mission builder and excellent horseman. Kino laid the foundation for the original San Xavier del Bac Mission. His lengthening the Camino Real into southeastern Arizona and Christianizing the natives served dual purposes of opening up accessibility to mining localities and the labor with which to mine metals at the base of Black Mountain in 1699. The Camino Real, known as the King's Highway, was a route traveled by conquistadors from Vera Cruz, Mexico, to Tucson. Spanish reports indicated that Padre Kino might have been involved in some of the earliest European silver-mining operations in the Santa Rita Mountains. It is not surprising that the Spanish followed the statement made by one of the world's foremost mining engineers, T.A. Rickard: "Trade follows the flag, but the flag follows the pick." (Courtesy of the author.)

In 1775, Capt. Juan Bautista de Anza II led an expedition of over 240 colonists and soldiers, including their family members with over 600 horses and mules and over 350 head of cattle, from the Mexican towns of Culiacán and Horcasitas to San Francisco, California. The objective of this venture covering over 2,700 miles was colonization and surveying potential sites for future mining operations. He owned several mining interests along the border in southeastern Arizona. (Painted by Fray Orsi, courtesy of the National Archives and Records Administration.)

This westward view of the boundary line between Arizona and Mexico in the heart of Nogales was taken in the late 1890s. Just a few miles south of this locality was the famous Planchas de Plata ("slabs of silver") discovery of 1736 near a ranch called Arizonac in northern Sonora, Mexico. (Courtesy of the National Archives and Records Administration.)

San Xavier Mission is the oldest European structure in Arizona, completed in 1797. It is rumored that $60,000 worth of silver utensils once adorned its altar. The silver came from mining operations in the neighboring Santa Rita Mountains operated by the padres at Tumacacori Mission. (Courtesy of Wendall Brown.)

Charles D. Poston, known as "the Father of Arizona," concentrated his mining efforts around the Tubac area. He formed the Sonora Exploring and Mining Company. Poston acquired funding from California and investors back east, including Samuel Colt. His influence in the region was phenomenal, since no established government existed in the area until the War Between the States. (Sketch by John R. Browne, courtesy of the National Archives and Records Administration.)

Military forts and detachments of armed soldiers were essential to protect mining ventures in southern Arizona from Apache deprivation led by raiders such as Geronimo, depicted at left. The remains of Fort Lowell in Tucson are pictured above. Established in March 1873, Fort Lowell operated just east of Tucson on Rillito Creek until being decommissioned in 1891. Because of its early appearance as a swarm of tents, it was originally known as Camp Lowell. However, soon after, local Tucson workers helped erect permanent adobe buildings, including officers' quarters, enlisted men's headquarters, an eight-room hospital, a guardhouse, a quartermaster's storehouse, a blacksmith shop, and a cavalry corral. After April 5, 1879, it was known as Fort Lowell and housed two to four companies of soldiers, contributing to the increased revenue in Tucson. (Above, courtesy of Wendall Brown; left, courtesy of Library of Congress Prints and Photographs Division.)

Between the years 1880 and 1930, it was commonplace for the company store in the mining towns of southeastern Arizona to issue trade tokens. These coins, often circular in shape, were made of aluminum, bronze, copper, nickel, and zinc. The tokens were limited in value, because they could only be exchanged for their stated value's worth of merchandise by the issuing store. Their assigned value ranged between 5¢ and $1. This Gleeson token has a reverse stating its value of 25¢ in merchandise not transferable. (Courtesy of Gleeson, Arizona, Museum.)

INCORPORATED UNDER THE LAWS OF ARIZONA

NUMBER 21 SHARES 75

MYSTERY MINING COMPANY

CAPITALIZATION $1,000,000

This Certifies that Mae Purcell Casey is the owner of Seventy five Shares of the Capital Stock of

MYSTERY MINING COMPANY

transferable only on the books of the Corporation by the holder hereof in person or by Attorney upon surrender of this Certificate properly endorsed.

In Witness Whereof, the said Corporation has caused this Certificate to be signed by its duly authorized officers and to be sealed with the Seal of the Corporation this 24 day of April A.D. 1923

SECRETARY PRESIDENT John Gleeson

SHARES $1.00 EACH

The stock certificate pictured here represents ownership of 75 shares of capital in the Mystery Mining Company. The signature at the bottom right denotes the town's namesake, John Gleeson. The assignee, May Purcell Casey, was the daughter of John Gleeson's sister. The Mystery Mine consisted of four patented claims worked intermittently between 1923 and 1942. The principal ores included lead, silver, and zinc. (Courtesy of Gleeson, Arizona, Museum.)

Arizona territorial governor Richard C. McCormick supported a mining-based economy similar to that of California. During his term in office between 1866 and 1868, he provided tax breaks to mining ventures throughout the state. (Courtesy of Library of Congress Prints and Photographs Division.)

Established in 1871, the San Carlos Indian Reservation served as a confine to secure Apaches who raided American and Mexican settlements, including mining towns and prospects. They were put to work on useful projects such as building irrigation ditches, which is depicted in the photograph below. Peridot Mesa on the San Carlos Indian Reservation is renowned for being the most productive locality for peridot in the world. Green peridot is faceted in jewelry stones. (Courtesy of the National Archives and Records Administration.)

Several officers and guests dine beneath a large saguaro cactus near Fort Thomas, located 39 miles southeast of the San Carlos Indian Reservation. Established on August 12, 1876, in honor of Maj. Gen. George W. Thomas, the fort was another army outpost used to provide stability for settlers and those who prospected and worked the mines between Globe and Morenci. (Courtesy of the National Archives and Records Administration.)

The Southern Pacific Railroad crew lays track near the future town of Benson. The town owed its existence to the railroad as both a transportation hub and a destination smelter for the mines near the Santa Rita Mountains. (Courtesy of the San Pedro Valley [SPV] Arts & Historical Society.)

The pusher engines seen here at the Benson roundhouse helped trains up the steep hills that surround the town of Benson. Pusher engines are secondary pull engines used to support the trains going up a steep gradient. After completing their assignment, they return to support the next train. As supporting engines, their maximum speed averages 52 miles per hour. They usually consist of older engines that are no longer in use for general operations. (Courtesy of SPV Arts & Historical Society.)

The arrival of the train was much heralded across southeastern Arizona. Families are gathered here at the old railroad depot in Benson to meet the arrival of travelers from either the eastern or western parts of the United States. Prior to 1918, when the US government took over the nation's express for the war effort, Wells Fargo & Company Express, as noted on the sign at far right, served as the country's first nationwide express company. Wells Fargo was committed to the transport of money, whether by stagecoach, steamship, railroad, pony rider, or telegraph. It was very much involved in both the transport and protection of gold shipments since its establishment by Henry Wells and William Fargo in 1852. (Courtesy of SPV Arts & Historical Society.)

The Benson roundhouse served the town of Benson as a major switching point for the railroad until it was moved to Tucson in the early 1920s. The diamond stack on the ground below was used by wood-burning locomotives to prevent sparks from escaping through the stack. Below is an image of the Southern Pacific Railroad roundhouse as it appeared in 1908. It contained eight stalls for servicing locomotives. Named in honor of Judge William A. Benson, a friend of Charles Crocker, who was the president of the Southern Pacific Railroad, Benson was selected as the transportation hub for the railroad to ship freight to and to obtain ore from the copper and silver mines in southeastern Arizona, including Bisbee, Contention City, Fairbank, and Tombstone. During the late 19th century, Benson rivaled Tucson as a rail transportation center. This was a boon to Benson's local economy, helping to establish hotels, saloons, and restaurants in the area. The impact also increased agricultural interests in the San Pedro Valley, many of which were established by Mormons. (Both, courtesy of SPV Arts & Historical Society.)

The railroad repair crew depicted here in the early 1880s is, from left to right, El Prieto, Necho, Lorenzo, Jordon, Merce, unidentified, and Steve Gonzales. These men contributed to the creation of the southern transcontinental railroad line. The railroad increased the population in and around Benson. In 1880, the population of Benson was around 300; however, by 1910 the town reached a population of 1,100. Benson was the temporary drop-off point for the 36-inch Rankin and Brayton furnace shipped from San Francisco on route to Bisbee. Smelting operations were established at the base of the Copper Queen Glory Hole. (Courtesy of SPV Arts & Historical Society.)

Here is the roundhouse in Douglas, Arizona, as it appeared in the early 1900s. The growth of mines and railroads were interrelated. This symbiotic relationship involved economic gain for both the railroad and the mining companies during the reliable transportation of ore to affordable processing facilities. (Courtesy of Library of Congress Prints and Photographs Division.)

$1 NOT TRANSFERABLE. $1

Pay to Esteven Romero

$1 In Merchandise Only at

Fiege & Co. Commissary and Meat Market

JOHNSON. - ARIZONA.

NO SE TRANSFIERE.

Bueno por $1 en Mercancias en la

Comisaria y Carniceria de Fiege y Cia.

F.M. Clifford

This early 1900s commissary note issued by a mining company in Johnson Camp, Arizona, to local miner Esteven Romero serves as an example of payment by script. Mining companies paid their employees in their own script signed by the mine boss endorsing the note's value for the said amount of goods at the company store, also known as the commissary. The script acted as credit advanced against the miner's wages. Below is a note issued by the Dragoon Mining Company to an employee who earned $1.25 per day. Orville Mickens (March 12, 1920–August 11, 2007), collector, curator, educator, historian, gunsmith, mechanic, prospector, and overall classic fellow who roamed the mountains and valleys of Southeastern Arizona in search of metals and relics, had a premier collection of mining script and tokens on display at his front-yard museum Frontier Relics in downtown Dos Cabezas. (Both, courtesy of SPV Arts & Historical Society.)

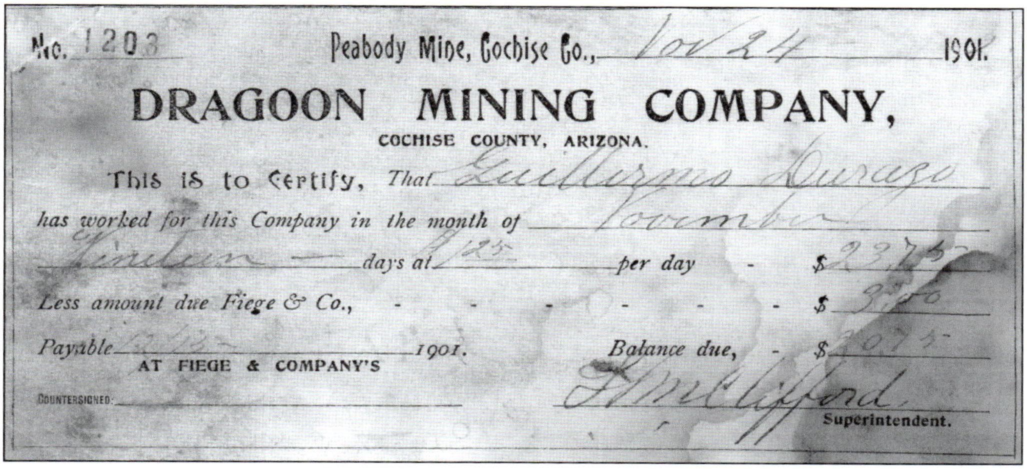

No. 1203 Peabody Mine, Cochise Co., Nov 24 1901.

DRAGOON MINING COMPANY,

COCHISE COUNTY, ARIZONA.

This is to Certify, That Guillermo Durazo

has worked for this Company in the month of November

Nineteen days at $1.25 per day - $23.75

Less amount due Fiege & Co., - - - - - - - $3.50

Payable 1901. Balance due, - $20.75

AT FIEGE & COMPANY'S

COUNTERSIGNED:

F.M. Clifford

Superintendent.

Many events, like hosting movies, were held at the Central Copper Company Community Hall in Dos Cabezas. The piano player at right appears to be taking a break from entertaining her audience, which includes the local town teacher, Earl Matteson, sitting fifth from left in the third row. His pupils are clearly seen in front, with the miners and their wives directly in the back of this happening venue. (Courtesy of SPV Arts & Historical Society.)

Located north of the upper power plant in Dos Cabezas, the Central Camp School hosted an average of 40 pupils. Earl Matteson, the teacher of this frontier school, appears with a stern face at the far right. It was his goal to instill a firm education in the children of the Dos Cabezas miners. (Courtesy of SPV Arts & Historical Society.)

Typists are seen at work at Old Mascot Copper Company's office in Dos Cabezas. By 1920, the Mascot Copper Company was enduring financial burdens and was absorbed by the Central Copper Company. (Courtesy of SPV Arts & Historical Society.)

Case's Place, a happening hub of activity in the town of Ruby named for Charlie Case, was a popular confectionery for the miners and sold cakes, pies, and chicken salad sandwiches from 1929 through 1939. Both Charlie Case and his wife, Forrest "Ma" Case, were the proprietors. The Ford Coupe with the rumble seat parked in front may have belonged to Walter Strange Pfrimmer. The guesthouse at the far right served mining officials visiting from out of town. Prefabricated houses, known as Ronstadt houses, stand at the far right. The horses probably belonged to local ranchers. (Courtesy of Tallia Pfrimmer Cahoon.)

The concrete jail built in 1934 probably operated through 1942, prior to Ruby's closing. The town's elevation is 4,219 feet above sea level, and Montana Peak, seen in the distance to the southeast, is 5,370 feet. The schoolhouse seen here is located south of the jail. The teacherage house is clearly seen between the school and the jail. (Courtesy of Tallia Pfrimmer Cahoon.)

This photograph depicts both town residents and visitors near Ruby. Hilda Allison Kinney and her husband, Lant Kinney, along with Julia Allison Edwards on mule and Jack Kelly holding a firecracker, pose for the camera in Ruby. Seen at the far right is a tent house, while the Dr. Jules Woodard house is seen in the far center background up on the hill. (Courtesy of Tallia Pfrimmer Cahoon.)

Two

Small Mining Towns and Localities

Southeastern Arizona contains a rich mining history involving an assortment of small mining towns and camps—the terms of which are used interchangeably. The lack of territorial authority in Arizona led to self-governing mining districts. Today, many have long since been abandoned, attaining the status of ghost towns, and the few that remain boast a sparse population. Several reasons factor into the fate of these sites, including diminishing supplies of ore and the lack of economic feasibility to continue mining operations. The average peak productivity of many of these towns and localities ranged between five and 10 years. Some examples of these towns are Alto, Courtland, Duquesne, Gleeson, Harshaw, Helvetia, Pearce, Ruby, Salero, Tiger, Tombstone, and Washington Camp. Some towns and localities remained in operation on and off again according to the ever volatile metal market, while others suffered the fate of flooding hindering the economic and accessible gain for future underground mining operations.

Collectively known as the Harshaw Mining District and Patagonia Mining District, the towns of Duquesne, Harshaw, Lochiel, Mowry, and Washington Camp were combined to create a productive mining region in southeastern Arizona from 1860 through 1972. Located near the Mexican border in the Santa Rita Mountains and Patagonia Mountains, these mining towns experienced their maximum output during the early 20th century, especially during the World War I years.

Located at the northwest end of the Santa Rita Mountains, the Helvetia camp consisted of small mines and towns. Miner Ben Hefti named Helvetia, or "liberty," for his native home of Switzerland. Located in the Santa Rita Mountains, it became an active mining locality in 1875, with over 20 recorded claims by 1883. Denoting its status as the preeminent Pima County mining camp, by the end of the 19th century, over 600 men were employed at Helvetia along with two stage lines that ran to and from Tucson. By 1911, a decline in copper prices forced the mines to close. With the advent of World War I, mining resumed; however, it never regained its earlier prominence. The demise of the town became evident when the post office closed in 1921 and the school two years later. Despite many years of sporadic mining operations after World War I, estimates conclude that around 17 million pounds of copper valued at $4 million were produced around Helvetia prior to 1950.

This 1909 photograph depicts the Alto camp and mill, located in the Santa Rita Mountains. Formerly known as El Plomo, meaning "lead" in Spanish, because of its abundant lead deposits, the mining community later changed its name to Alto, meaning "high" in Spanish. The Spanish discovered gold at this site in 1687. The early 20th century saw the town reach a population of several hundred. A post office was active nearby from 1907 to 1933. (Courtesy of USGS.)

Depicted looking southwest in this photograph is a view of the Bonanza mine located near Duquesne in 1909. Owned by the Duquesne Mining and Reduction Company, its workings were dug below 650 feet, and its equipment consisted of a 50-horsepower electric hoist and an Ingersoll compressor. (Courtesy of USGS.)

Perhaps one of the most colorful names for a southeastern Arizona mining camp is the Total Wreck Mine, located in the northeast part of the Empire Mountains. The silver mine got its name from a ranch employee of Walter L. Vail, John T. Dillion, who upon surveying the boulder-strewn property of granite and quartz formations on the eastern foothills of the Empire Mountains declared the area a total wreck. This image depicts the mill found at the site, built in 1883. (Courtesy of USGS.)

Here is the Camp Washington and Duquesne reduction plant, located on the eastern slope of the Patagonia Mountains, as it appeared in 1909. Although mining operations were conducted here in the 1860s, hostile Apache Indian raids hindered full development until the late 1880s, when a smelter, reduction plant, offices, boardinghouse, and stores were established along with a population of 2,000 in the area. The Duquesne Mining and Reduction Company made its headquarters in Duquesne with a neighboring reduction plant a mile north. The inventor of the railroad air brake, George Westinghouse of Westinghouse Electric Company, owned not only property with a vast collection of mining claims in Duquesne but also a house that had modern plumbing with hot and cold water. (Courtesy of USGS.)

Found on the eastern slope of the Patagonia Mountains are two mining camps, Washington and Duquesne, which are situated three quarters of a mile apart. Washington Camp and nearby Duquesne prospered in the 1890s. Located five miles south of the Mowry Mine and 17 miles east of Nogales, Washington Camp had a general store, miners' residences, and a school. This photograph, taken looking west in Washington Gulch, depicts the Washington Camp and Duquesne Reduction Plant in 1909 with the Pride of the West Mine at left and the Patagonia Mountains in the background. (Courtesy of USGS.)

Located near Patagonia in Santa Cruz County, the World's Fair Mine was owned and operated by the Commonwealth Development Company, which also owned the Commonwealth Mine in Pearce. Consisting of 13 claims, the World's Fair Mine had a 600-foot deep-mine shaft with several miles of workings. The principal metals mined at this locality included silver-lead ore. The years of production ran from 1879 through 1954, with a product valued at over $1.6 million. (Courtesy of USGS.)

A view along Sonoita Creek, with the town of Patagonia in the foreground and the Santa Rita Mountains in the background, is surrounded by a plethora of mining localities. Mount Wrightson, the highest point in the Santa Rita Mountains at 9,453 feet, is evident on the far right. Both Mount Wrightson and nearby Mount Hopkins were named for miners employed by the Salero Mining Company. They met their fate from marauding Apaches. (Courtesy of USGS.)

The Mowry Mine, seen above, boasted a large lead-silver deposit in the Patagonia Mountains. Originally worked by the Jesuits during the Spanish era, the mine was acquired by Sylvester Mowry in 1859. It purportedly supported the war effort of the Confederacy by supplying lead bullets for its soldiers during the War Between the States. During the Mowry Mine's peak production, it cleared over $4,000 per week. The lead-silver mine closed after the War Between the States. (Courtesy of USGS.)

The US government arrested its proprietor, an ex-officer from Fort Yuma named Sylvester Mowry, for being a Confederate sympathizer. Although the charges were dropped, he never regained his mining operations in southeastern Arizona. He died in 1871 in England while trying to revive both his health and his fortune. A smelter was built at the site in 1910, producing 5,000 tons of slag. (Sketch by John R. Browne, courtesy of the National Archives and Records Administration.)

This 1909 photograph is looking north-northwest at the Copper World Mine and Mill from the vantage point of Isle Royal in the Santa Rita Mountains. Developed at an elevation of 4,850 feet in 1900, the mine produced copper and iron sulfides and copper carbonates, along with a small amount of gold and silver. (Courtesy of USGS.)

This 1909 view is of the Helvetia mining camp. In the background are the Santa Rita Mountains, where mineral deposits were discovered in limestone. This photograph includes, from left to right, the mines Heavy Weight, Copper World, Mohawk, Leader, Isle Royal, Old Dick, and Omega. The first decade of the 20th century was the high-water mark for Helvetia. The area boasted four saloons; several stores, including a meat market, Chinese laundry, and barbershop; and a school. Adobe ruins, scattered mine tailings, and a small community cemetery are all that remain of Helvetia, which is located on the western base of the Santa Rita Mountains. (Courtesy of USGS.)

The Jesuits and their Mexican labor originally worked Salero in the 18th century. This image was taken in 1909. The name Salero derives from the Spanish term for a saltcellar. Legend has it that silver mined at the site was used to design the saltcellar in honor of the bishop of Sonora who was visiting the nearby Tumacocari Mission. Located on the southwest slope of the Santa Rita Mountains, the property is at an elevation of 4,150 feet. (Courtesy of USGS.)

This 1909 image of the Sweet Mine and smelter plant, located six miles north of Patagonia, denotes vein outcroppings in the foreground. The Mansfield Mining & Smelting Company dug a 360-foot shaft, which produced a small amount of ferruginous copper ore. (Courtesy of USGS.)

Located 19 miles east of Tombstone, the once-booming town of Courtland, which boasted over 2,000 inhabitants, is now entirely abandoned. It was named after Courtland Young, the brother of W.J. Young, who owned the Great Western Mining Company and had a vested interest in the town of Courtland. Railroad spurs were built to the mining towns of Courtland and Gleeson in 1909. The ore was shipped to the smelter in Douglas. Production at these localities was greatly reduced by 1933, and as a result, the railroad tracks were dismantled. (Courtesy of Gleeson, Arizona, Museum.)

The Soto Brothers & Charles Renaud store is at the heart of the old mining town of Pearce, located 28 miles south of Willcox. Jimmie Pearce, a Cornishman and Tombstone miner and rancher, traveled around the future site of Pearce in 1894. While idly breaking a piece of quartz with a hammer, Pearce discovered a piece of gold. Between February and July 1895, he staked out 11 claims in the area, dividing each of them with his family members. Jimmie Pearce named his discovery the Commonwealth. Several years later, Pearce sold the mine for $250,000 in cash to a group of financiers forming the Commonwealth Mining and Milling Company. The mine's closure in the 1930s caused the business district to decline, and by the 1960s, all that remained were a post office, country store, and museum. (Photograph by Kevin Bays, courtesy of the author.)

Named after Irish immigrant John Gleeson, who settled in Pearce in the 1890s, Gleeson is located 16 miles east of Tombstone. The Silver Bell Mine yielded rich silver-lead ore in the years prior to World War I. The site was originally named Turquoise after the abundance of turquoise found in the area. The demand for copper accelerated Gleeson's economy. The town boasted a population of 500 by 1910; however, a devastating fire in 1912 destroyed 28 buildings in town. Afterwards, the town was rebuilt. The demise of Gleeson by 1940 was due to the declining price of copper. The image above dates to 1900 and includes three-year-old Leona "Nona" Langley, held by her father Noah Langley on his knee. (Courtesy of Richard Langley Ware.)

Depicted in this early-1900s photograph are horses hauling tubular boilers to the Commonwealth Mine near Pearce, Arizona. Cylindrical boilers consist of a series of long internal tubes running horizontally the length of the boiler, surrounded by water. These boilers were positioned above a fire, sending heat and smoke through the long tubes, thus heating the surrounding water and creating steam. Released through the smokestack, the steam was used to power mining and milling machinery. (Courtesy of Richard Langley Ware.)

Work at the Collins Cut stopped at 1,125 feet due to the rising levels of water. The Mohawk Mine and its 10-stamp mill were established in the 1890s, with production beginning in 1896 at 70 tons per day. The Mammoth Mine furnished many cabinet specimens of the lead mineral wulfenite during its operations. (Courtesy of the Arizona-Sonora Desert Museum.)

George Biggs, a carpenter and handyman in Mammoth, stands at left by several miners employed at the Sunrise Shaft at Tiger. Although little remains of the mining camp of Tiger, located three miles southwest of the current town of Mammoth, its reputation as a premier site for rare and beautiful minerals is firmly established in the history of mining in Arizona. Gold, molybdenum, vanadium, lead, zinc, and silver are some of the 89 mineral species that were found at this locality. The first mining claim established at the site dated back to 1879. Multiple shafts and mills were erected at the site. Production declined after World War II due to competition from international lead and zinc markets and declining metal prices. Flooding in the mine shafts also hindered operations. By 1954, the post office closed, signaling an end to the townsite. (Courtesy of the Arizona-Sonora Desert Museum.)

Originally established as Schultz in 1894, after an Austrian immigrant named Frank Schultz, the mine was later renamed Mammoth because of the sizable gold deposit believed to be of mammoth proportions. The mine is also referred to as the Mammoth-St. Anthony mine after a nearby catholic church by the name of St. Anthony. The mine and town was later purchased by Sam Houghton, who renamed them "Tiger" after his college mascot, the Princeton Tiger.(Courtesy of the Arizona-Sonora Desert Museum.)

As it appeared in 1890, the Mammoth Mill in the town of Mammoth processed ore delivered in tramway buckets from Tiger. The town had a population of 700 and boasted six saloons. By 1900, the Mammoth Mill consisted of 70 stamps weighing 820 pounds each. The vibrations were audible over a distance of three miles. (Courtesy of the Town of Mammoth.)

The town of Ruby was located in the Oro Blanco District, which is often referred to as the Ruby District; it was noted for its deposits of lead and zinc. The Spanish originally mined the area for gold and silver. Ruby includes the Montana Mine, a renowned late-19th-century-type locality for surface gold and later its abundance of lead, zinc, and copper. Named in 1912 for Lillie Ruby, the wife of the town's postmaster and storekeeper Julius Andrews, it once boasted a population of 1,200. (Courtesy of Tallia Pfrimmer Cahoon.)

Walter Strange Pfrimmer is shown hatless among a surveying party of seven in the hills of Ruby prior to 1930. He served as a mining engineer for the Eagle Picher Mining and Smelting Company from 1926 until 1938. A tripod with a transit stands in front of the iron contraption in the background, which is a surveying auger drill, used to set telegraph poles. (Courtesy of Tallia Pfrimmer Cahoon.)

Deep in a Ruby mine shaft, Walter Strange Pfrimmer surveys the mining operation with his foot on the bench. Inscriptions on the rock wall depict names, initials, figures, and dates. This photograph postdates the inscription of 1929 on the wall by several years. A level is clearly seen leaning to the side on the left, while an ore cart and chute are clearly seen on the right-hand side. (Courtesy of Tallia Pfrimmer Cahoon.)

A canvas bag with leather straps containing a compass and metal writing pad is evident to the left. The electric trolley at right was used to haul ore. (Courtesy of Tallia Pfrimmer Cahoon.)

These timbers, found in Ruby's underground workings, measure 12 to 15 feet in height. They were imported to Arizona and distributed through Amado, Arizona. Walter Strange Pfrimmer, with hat and light shirt, sits with a prospect pick at his side and enjoys some idle time with his fellow miners beneath carbide lamps. A shovel and crowbar used for prying are evident in the background. (Courtesy of Tallia Pfrimmer Cahoon.)

Walter Strange Pfrimmer sits with a carbide lamp and camera case on his hip. Walter graduated from the Colorado School of Mines as a mining engineer. Two sizes of carbide lamps used in mining were a larger hand-held version or the smaller size lamp attached to the miner's helmet. (Courtesy of Tallia Pfrimmer Cahoon.)

Walter Strange Pfrimmer stands with a hat on at the far left. The blanket on the horse is known as a serape. Developed by the Spaniards in the late 16th century in northern Mexico, the serape provides protection from the environment in case of cold and hot temperatures. Serape is the English word derived from sarape, which was used to describe a Mexican shawl or cape worn by men. (Courtesy of Tallia Pfrimmer Cahoon.)

Montana Peak is clearly seen looking from west to east in this photograph dating back to 1934. A small Ronstadt (manufactured home) is seen at center in the lower portion of the photograph, while a display of quartz rock is evident on the hill. A big tent is located at the far right. The Montana Mine was the leading producer of lead and zinc in Arizona during the late 1930s. It ranked third in output of silver in Arizona in 1938. (Courtesy of Tallia Pfrimmer Cahoon.)

With both its hoist and mill evident, the Montana Mine, or Ruby Mine as it was sometimes called, is clearly seen across Ruby Lake. A cooling tower sits on the hill behind the mine building. George Cheney built Ruby Lake in the early to mid-1890s. Ruby Lake is 25 to 40 feet deep, depending on water runoff. (Courtesy of Tallia Pfrimmer Cahoon.)

This Upper Tailing Dam photograph, taken on February 16, 1931, depicts a view of Ruby Lake looking toward the west. The apparatus emptying water is maintaining a constant water level for the lake to prevent overflow. The first store built in Ruby appears across Ruby Lake. (Courtesy of Tallia Pfrimmer Cahoon.)

This photograph, taken prior to 1934, is looking east at the town of Ruby with the waters of Ruby Lake in the distance. The mill is to the left of the hoist, and the mine pad is at the right. Bunkhouses are located on the left-hand side. Dr. Jules Woodard's hospital is at the far left behind the eave of the bunkhouse. The company office is located behind Dr. Woodard's hospital. The assay office on the hill is next to the cooling towers behind the mill. In 1928, the Eagle-Picher Lead Company built a mill that processed 250 tons of ore on a daily basis. Several hundred miners were employed at the mine. By 1941, mining operations ceased at Ruby. Today, at least a dozen buildings remain standing on the town's premises. (Courtesy of Tallia Pfrimmer Cahoon.)

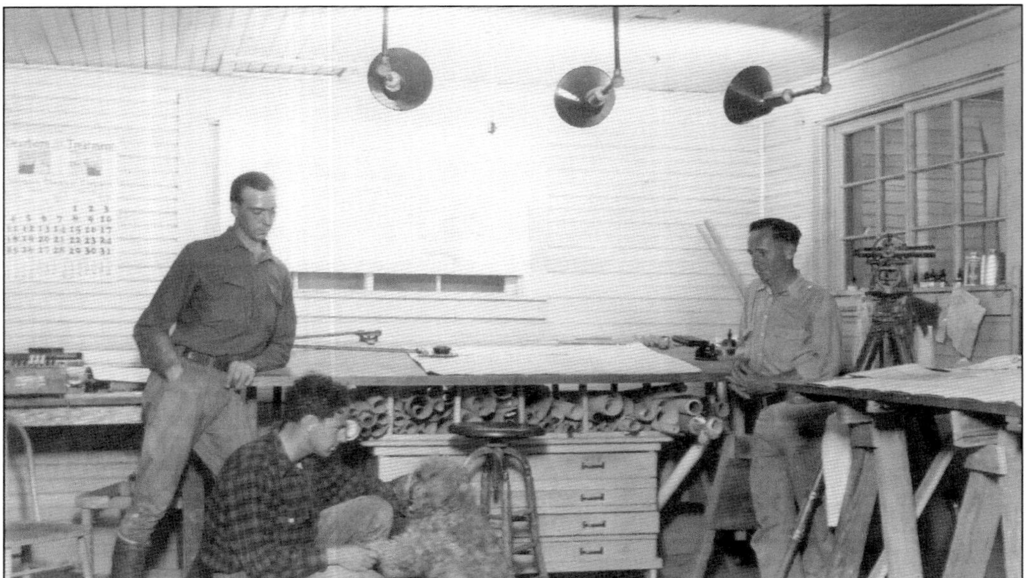

The office of the Eagle-Picher Lead Company in Ruby is a hub of activity for these miners with an attentive dog. The two men on the left are unidentified, but the one on the right is Walter Strange Pfrimmer. According to the background calendar, the date appears to be May 1930. (Courtesy of Tallia Pfrimmer Cahoon.)

This image is looking southwest toward the mill and the hoist. Built in 1928, the mill had a daily capacity of processing 250 tons of ore. Ruby Lake and water from Gold Boulder Dam southwest of Ruby supplied water for the mining and milling operations. The outcropping of quartz is clearly evident on the upper hill at left. During the 1870s, American prospectors discovered the prominent quartz outcrop that later became known as the Montana vein. Immediately, mills were built to treat the rich ore deposits in the area. (Courtesy of Tallia Pfrimmer Cahoon.)

The upper power plant belonging to the Central Copper Company appears to be a happening place in Dos Cabezas. The five individuals posing for the camera are probably plant managers and employees. Named after the Spanish for "two heads," applied to the 8,354-foot twin peaks, the mining town of Dos Cabezas flourished between 1880 and 1927. (Courtesy of SPV Arts & Historical Society.)

During the town's early years, the ore was originally transported by truck and later by rail to the nearest railhead in Willcox and was then transported to the smelter in Douglas, Arizona, or El Paso, Texas. The track belonging to the Mascot & Western Railroad is clearly defined below the noisy truck known as the "Black Cat" or "Doodle Bug." The truck was converted to haul supplies, passengers, and mail twice a day on the track that ran between Dos Cabezas and Willcox. Lon Perkins, the driver, stands at left, while Lon Haley, the conductor, is in the middle. The third man at far right is unidentified. The buildings in the back are storage sheds for the Mascot Copper Company. Despite scattered mineralization and deposits of low-grade ore, the town managed to hold itself together, having produced 7,000 tons of gold, silver, and lead by the first decade of the 20th century. It was also during this time that the Mascot Copper Company was formed to capitalize on a gold-bearing quartz dike in the area. Veins of high-grade chalcopyrite were removed during the creation of the Consolidated Tunnel, which was sunk over 1,150 feet in depth. (Courtesy of SPV Arts & Historical Society.)

Three

Principal Mining Towns and Localities

Several principal mining localities exist in southeastern Arizona. Examples include Bisbee, Globe, Miami, Mission, Morenci, Ray, Superior, and Silver Bell. Bisbee, located 90 miles southeast of Tucson in the Mule Mountains, owed its initial discovery to a detachment of prospecting cavalry scouts under the command of Lt. Jack Dunn from Fort Huachuca. They arrived in the Mule Mountains in the fall of 1877 and began staking claims in Tombstone Canyon because of the rich deposits of copper, lead, and silver in the area.

Established as a mining camp in 1880, Bisbee was named after Judge DeWitt Bisbee, who financed the Copper Queen Mine. A rich copper carbonate ore body was discovered that same year in the form of the Copper Queen claim. Several smelters were built in Bisbee, and by 1890, most of the trees in Bisbee had been cut down for use in maintaining the smelter. By the turn of the century, Bisbee was the largest city between St. Louis and San Francisco, with a population of 20,000 people. During its prime in the early 1900s, it had over 45 active saloons on Brewery Gulch and was considered the hot spot for entertainment between El Paso and San Francisco.

In 1917, the town of Bisbee was the scene of a major labor strike that resulted in what became known as the Bisbee Deportation. Labor disputes between miners and companies over working conditions and wages became heated as the price of copper soared because of the US military's need for the mineral in World War I. Labor saw an opportunity to strike to make their voices heard. This wartime uprising angered both Phelps Dodge and the townspeople of Bisbee. On July 12, 1917, several thousand labor union members were rounded up on boxcars by an armed vigilante force in Bisbee, Arizona, and deported to Columbus, New Mexico.

Several prominent open-pit mining operations, including both the Sacramento Pit and later the Lavender Pit, developed in Bisbee during the 20th century. The Lavender Pit, named in honor of Harrison Lavender, a former manager of Phelps Dodge's Bisbee mining operations, was developed in the mid-1950s and engulfed the Sacramento Pit. Over 351 million tons of material was extracted during its development. Mining ended at the Lavender Pit on December 14, 1974. By the time major mining operations in Bisbee ceased in 1975, an estimated 3 million ounces of gold, 97 million ounces of silver, and 8 billion pounds of copper, along with over 600 pounds of both lead and zinc, made for a 100-year total of over $6 billion in metals extraction. Today, the 300-acre, one mile long and a half a mile wide, 950-foot-deep pit serves as a tourist attraction. Since its closure, the Lavender Pit continues to be a site of copper recovery through a process known as chemical heap leaching, wherein low-grade copper ore from the excavated waste rock is successfully extracted using ferric sulfate and sulfuric acid solutions.

This 1902 view of the town of Bisbee from Sacramento Hill was taken in the Mule Mountains. The Holbrook shaft is seen in the foreground, and the Czar shaft is evident in the middle. The Mule Mountains average 21 miles in length and 33 miles wide, and the highest elevation in these mountains, Mount Ballard, is about 7,400 feet. These granitic mountains acquired their name from the Mexicans, who called that locality Puerto de las Mulas, or "mule pass," because nearby springs provided water for both men and their pack animals. (Courtesy of USGS.)

A view of the Lavender Pit from behind a fence shows it as it appears today. (Courtesy of the author.)

The Irish Mag and Spray shafts are seen from the vantage point of Sacramento Hill on December 19, 1902. The Spray shaft, owned by the Copper Queen Consolidated Mining Company, appears on the right at 954 feet in depth. Originating in 1889, the Spray shaft provided 25 years worth of steady oxide ore production. Owned by the Calumet and Arizona Mining Company, the Irish Mag shaft attained a depth of 1,200 feet by 1902. Named after a Bisbee dance hall girl, the Irish Mag Mine yielded the discovery of a large high-grade copper ore body buried under layers of limestone at the far end of Brewery Gulch in 1899. The rich grade of the oxide ore averaged 25 percent while the sulfide ore averaged 12 percent. A year later, the Calumet and Arizona Mining Company selected Douglas as the site to build its smelter to treat the copper ore from the Irish Mag. The smelter turned out thirty tons of copper daily. The Irish Mag claim provided over $15 million in dividends until its closure after 11 years of operation. (Courtesy of USGS)

A major expense in mining is the transport of equipment and material back and forth from the mines and then the ore to the smelters or markets. The railroad in the town of Fairbank provided the shortest link for Bisbee to connect to the Sonora Railroad between Benson and Nogales. A toll road was also built over Mule Pass to Fairbank, and by 1884, commerce was transported along this route by mule teams, hauling an average of $7.25 per ton of copper ore. Due to a lack of coal and timber in southeastern Arizona essential to make charcoal for running the smelter, coke from England and Trinidad, Colorado was brought to Bisbee using 20-mule team wagons. (Both, courtesy of SPV Arts & Historical Society.)

This 1918 image from Bisbee depicts miners extracting manganese ore from the surrounding Naco limestone. (Courtesy of F.L. Ransome, USGS.)

A 1902 view of Sacramento Hill, at left center, depicts the multiple mining operations around Bisbee, including the Lowell Mine at center, the Irish Mag Mine at right center, the Spray Mine at the far right, and the Holbrooke Mine at the lower right. Sacramento Hill, located southeast of Bisbee, became the focal point of a new type of mining in 1917. The Sacramento Pit became renowned as the first open-pit copper mine in Arizona. Steam shovel work began at Sacramento Hill in Bisbee in 1918. Over 34 million tons of earth and rock were removed from its depths by 1929. (Courtesy of F.L. Ransome, USGS.)

The Benson Smelter operated between 1882 and 1885, processing several thousand tons of ore from the Flux Mine in the northern end of the Patagonia Mountains. Although Benson was not itself a mining town, it possessed a small smelter that accommodated ore from various mining localities in both the Santa Rita and Patagonia Mountains. Later, the smelter was shipped to Mexico. (Courtesy of the San Pedro Historical Society.)

The smelter stacks are clearly visible in this 1940s photograph of the Douglas Reduction Works. Founded in 1901 as a smelter town, Douglas was designed to accommodate the massive deposits of ore mined at Bisbee. Located 26 miles east of Bisbee, Douglas had two smelters, with one being built by the Calumet and Arizona Mining Company in 1902 and the other by the Copper Queen in 1904. The latter was in operation until 1931, and the former was acquired by the Phelps Dodge Corporation upon its takeover of the Calumet and Arizona Mining Company. Smelting operations ceased at the site in 1987, and the smoke stacks of the smelter were demolished on January 13, 1991. (Courtesy of Library of Congress Prints and Photographs Division.)

The arrival of the railroad in Bisbee in 1889 was a great event. Engine No. 1, seen in this photograph, dates back to the War Between the States. It hauled logs in the Great Lakes region and the Pacific Northwest prior to its service transporting copper ore from Bisbee to Fairbank or to El Paso, Texas. In 1902, the Phelps Dodge Corporation, with the resources of Dr. James Douglas, who owned the subsidiary Copper Queen Company, built the El Paso & Southwestern Railroad, connecting Bisbee directly to El Paso. For its many years in the service of the El Paso & Southwestern Railroad, it was saved from destruction and now resides at Union Plaza Transit Terminal in downtown El Paso. Dr. James Douglas (1837–1918), the renowned metallurgist, chemical geologist, and mining engineer who developed the mining operations in Bisbee during its first several decades of operation, established and ran over 400 miles of railroad track while serving as president of the El Paso & Southwestern Railroad, the Nacazari Railroad, and the Morenci Southern Railroad. Douglas provided innovations to the mining industry, such as refined methods of calcining ores and advanced ways to improve the extraction and processing of copper. By the early 1920s, over 1,000 miles of track served the interest of Phelps Dodge & Company. (Courtesy of SPV Arts & Historical Society.)

Depicted in this 1898 photograph is a view of the reduction works and copper mines of Globe. Globe may have been named after the discovery at a local mine of a circular chunk of silver embedded with lines similar to the continents of the Earth. It may also be attributed to a large circular silver boulder in the area. (Courtesy of the National Archives and Records Administration.)

The shaft of the Old Dominion Mine in Globe, Arizona, dates back to the 1880s and operated until 1931. The property, originally developed by the Lewisohn brothers of New York, included both a railroad and a smelter. Owned by the Old Dominion Copper Mining Company until the mine eventually extended to a depth of several thousand feet, its most prosperous years occurred under the ownership of Phelps Dodge after 1903. (Courtesy of USGS.)

This 1920s image of the Grey shaft shows the 75-horsepower electric hoist seen in the foreground. Owned by the United Globe Mines Company and connected to the 1,200-foot level of the Old Dominion Mine, its mill and smelter are clearly defined in the background. The Old Dominion Mine became one of the richest copper producers in the district, producing over $134 million in gold, silver, and copper. (Courtesy of USGS.)

A view of the Buffalo Mine from the nearby Old Dominion Mine, close to Globe, includes the incline where ore cars were lowered to the bins down below for transport by railroad. Aerial mine tramways acted as a cost-efficient means of transporting ore and supplies in rugged, inhospitable terrain that would hinder surface transportation. They operated on heavy stationary cables stretched between terminals to form a track. The Buffalo Mine had its own smelter that operated in 1882. (Courtesy of USGS.)

A view of the Miami District from the south in 1911 includes the Miami High School in the foreground, the mill and concentrate bins of Miami in the far distance, and the Miami power plant and International smelter at right. Miami, or "Mima" as it was pronounced, was named after Mima Tune, the fiancée of "Black Jack" Newman, a miner who located the Mima Mine. A reduction plant was built in 1909, and the town later became known as Concentrator City. (Courtesy of USGS.)

The Shannon Copper Company holdings at Morenci were purchased by Phelps Dodge in 1919. This view of the Shannon incline was captured from the bottom of Chase Creek. The Coronado Railroad, as it was known, ran from the bottom of the canyon to Clifton. The incline shown here was one the seven inclines used to transport ore from the top of the hill to the ore bins in the canyon for further shipment by rail. (Courtesy of USGS.)

The top of the Shannon incline, located at the summit of Metcalf Mountain 800 feet above the town of Metcalf, was named after the Shannon Copper Company, established in 1899. It remained in existence until the Arizona Copper Company bought it out 20 years later. (Courtesy of USGS.)

This view of the downtown of Clifton includes the smelter run by the Arizona Copper Company. The smelter was located at the confluence of the San Francisco River and Chase Creek, and the geography of the area includes a mixture of basalt, rhyolite, and granite. Discovered by Lt. John G. Bourke while on a scouting mission with the US Army in 1869, the rich copper deposits around the area that would eventually be called Clifton were not exploited until the arrival of a group of prospectors and miners from Silver City, New Mexico, in 1872. The town was probably named "Cliff Town" after the towering cliffs in the immediate vicinity. It may also have been named after Henry Clifton, who prospected in the region in 1864. (Courtesy of USGS.)

Here is a 1901 view of the operation at the top of the Shannon Mine. The miner is directing a mule train of ore. The town of Clifton, along Chase Creek in Greenlee County, began at the site of the Leszinsky brothers' adobe smelter in the 1870s. By 1882, the Arizona Copper Company acquired the claims of the Leszinsky brothers. The copper ore was rumored to have been so abundant that it could be quarried rather than mined. (Courtesy of the Greenlee County Historical Society.)

The Coronado Railroad is shown running by the engine house at Metcalf. There is a look of pride among the men who ride the baby-gauge locomotive. The baby-gauge (20-inch) railroad was built in 1879, connecting Metcalf and Clifton. Phelps Dodge entered the mining business in Arizona in 1881 and assumed full ownership of the Detroit Copper Mining Company by 1917. In 1921, Phelps Dodge acquired the Arizona Copper Company, which gave it control over the mining operations at Morenci and Metcalf. (Courtesy of the Greenlee County Historical Society.)

Miners are loading ore onto cars in the granite porphyry deposits found in the open cut on the first level of Metcalf mines during the late 19th century. During the next 50 years, various companies, like the Arizona Copper Company, Detroit Copper Company, and ultimately Phelps Dodge, financed the growth of what became known as the Clifton-Morenci area. Later, an open-pit mine was developed on the site that once was Metcalf. Today, it is the largest open-pit porphyry copper mine in North America. (Courtesy of USGS.)

This photograph, produced in June 1901, provides a view of the Old Dominion Mine and smelter. Although flooding proved a challenge during its early years of operation, the mine was later acquired by Phelps Dodge from the Lewisohn brothers of New York. Phelps Dodge provided the capital to modernize equipment and overcome the flooding issues. It ceased operation during the Great Depression. (Courtesy of USGS.)

Prospecting around the future Globe-Miami District began in 1874. By 1882, copper exploration led to the development of the Old Dominion and Buffalo veins along with the United Globe Mine deposits, which were originally established with the hopes of unearthing a silver bonanza. In 1886, the town of Globe claimed that all US copper coinage employed copper from its locality in its mintage. The mining town of Miami, located seven miles west of Globe, gained prominence after the first decade of the 20th century, when copper deposits were discovered at the base of the Pinal Mountains. This image includes an aerial view of the surface equipment found at the Inspiration Consolidated Copper Company around Miami. (Courtesy of USGS.)

The town of Superior owed its existence to the Magma underground mine operated by the Magma Copper Corporation. Depicted here is the town of Superior near the rugged cliffs of Apache Leap. This landmark acquired its name from the actions of Apache Indians when provoked by the US Army. Spotted by a heliograph operator on Picket Post Mountain, the base camp of the Apache was revealed to American troops, who cornered 75 Indians in the said locality. Forced to either surrender or fight to the death, the Apaches leaped from the high cliffs to their death rather than face their opponent in combat. (Courtesy of the author.)

This northwest view of the Ray District from the east side of Mineral Creek was taken in 1910. The No. 1 Mine operation is evident across the creek at the far left. The workings of the Ray Consolidated Copper Company are clearly seen, including the head frame, crusher house, and ore bin belonging to the No. 2 Mine. Behind the tall cactus are the features of the No. 3 Mine. Teapot Mountain, the prominent landmark at Ray, appears in the distance. (Courtesy of USGS.)

In 1912, American Smelting and Refining Company (ASARCO)'s Hayden smelter began operations to process ore from the Ray Mine. Hayden was founded as a company town that year near the confluence of the Gila and San Pedro Rivers. Its 1,000-foot-tall chimney, built in 1974 as part of the Hayden Copper Smelter, is the tallest free-standing structure in Arizona. The smokestack disperses residual gases. The lime and filter plant facilities are still actively operated. (Courtesy of the author.)

The history of the Ray Mine dates back to the establishment of the Mineral Creek Mining Company in 1880. The Arizona Hercules Copper Company built the town of Ray in 1909. Originally located 64 miles east of Phoenix in the Dripping Springs Mountains, the town no longer exists, having been consumed by the open-pit mine. British financiers and mining engineers influenced the expansion of the copper mine at Ray into a large-scale operation that eventually evolved from underground workings to open-pit by the early 1950s under the ownership of the Kennecott Copper Corporation. An 85-ton overhead-catenary-powered electric locomotive built by General Electric in 1941 worked both at Kennecott Copper's Chino Mines Division in New Mexico and in Ray, Arizona. (Courtesy of Patty Sirls.)

Teapot Mountain provides an amazing backdrop to the town of Ray. Miners' homes are evident in the background, along with a gallows frame (or gallus frame) on the hillside. The crushers are clearly seen above the ore bins, while the railroad tracks parallel the mineral creek wash in the foreground. (Courtesy of Ken Shake.)

A 1920 view of the Senator Morgan Mine in Twin Buttes shows the Sierrita Mountains from the south. There is an ore zone along the low ridge, and a trestle bridge is evident from the hoist. On the extreme left of the hoist is part of the dump. The mine included 500 feet of working shafts with a maximum depth of 150 feet. In the distance are Helmet Peak and the neighboring limestone hills near the San Xavier and Mineral Hill Mines. Freeport-McMoRan bought the long-closed Twin Buttes Mine site for $200 million as of January 2010. (Courtesy of USGS.)

A photograph from above of the open-pit mines 15 miles south of Tucson shows the Asarco-Mission Complex on the left, the Twin Buttes Mine in the upper right, and the Sierrita Mine in the lower right with its adjacent tailings pond. Known for its copper and molybdenum deposits, the Asarco-Mission Complex produced 53,700 tons of ore per day during 2008. The Sierrita Mine includes the same mineral deposits with the inclusion of the rare metal rhenium, which is used for high-temperature alloys, aircraft engine parts, and lead-free gasoline. According to ASARCO (American Smelting and Refining Company), over six times the amount of earth moved to dig the Panama Canal has been mined at this locality. Sierrita Mine has been dug a quarter-mile deep and spans a distance of over two miles from north to south and over one and three-quarter miles from east to west. (Courtesy of NASA.)

The smelter at Sasco (Southern Arizona Smelting Company), located five miles north of Silverbell, was built in 1907 and had several 350-ton blast furnaces and converters that processed 800 tons of ore daily from the Silver Bell Mountains. Declining copper prices and depletion of high-grade ore forced its closure several years later, and it was eventually dismantled in the 1920s. The nearby railroad was dismantled and sold by 1933. The community at one time contained over 600 residents. A site long abandoned, it is noted for colorful slag located near the remains of the smelter. Slag is defined as a dense glassy material composed of iron and silica, otherwise known as an iron silicate. (Courtesy of Albertype Company.)

The smelter stack base and ruins remain standing at Sasco as of January 28, 2011. The ultimate failure of the Sasco smelter was due to stagnant copper prices and the overextension of both the holdings of the Imperial Copper Company and later ASARCO. (Courtesy of the author.)

Located one mile east of the Sasco smelter are the basalt block remains of the now-roofless Rockland Hotel. The hotel belonged to the Sasco townsite and accommodated employees at the smelter. (Courtesy of the author.)

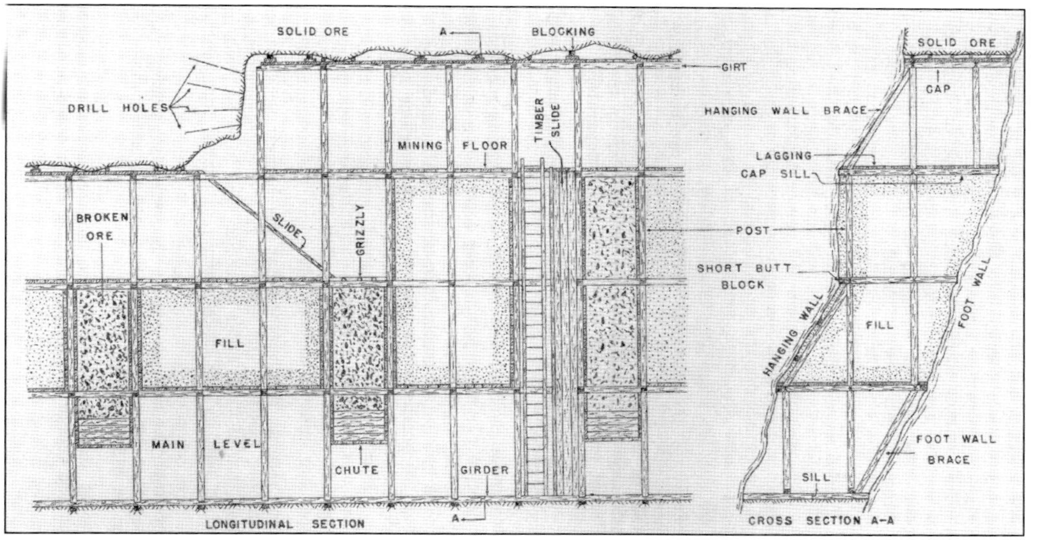

"Cousin Jacks," or Cornish miners, whose history dated back to the tin mines of southern England, contributed to the development of mining technology in southeastern Arizona. One of these advancements was the creation of dewatering pumps, whose purpose was to remove water from underground workings. They also developed a method called square-set stoping, a form of mining that added additional support to underground workings through the use of reinforced timber. This was helpful to underground miners in mining towns throughout southeastern Arizona because of the heavy ground that necessitated additional support. The daily wages for 10- to 12-hour shifts during the spring of 1909 ranged from $3.50 to $5 for foremen, $4.50 for engineers, $4.25 for crane men and machinists, $4 for steam locomotive engineers, and $2.40 for laborers. (Courtesy of Arizona Board of Regents for the University of Arizona.)

Evident in this photograph is an example of square-set stoping. A hoist is located nearby, and a miner is holding a carbide lamp on hand beneath reinforced timber. (Courtesy of SPV Arts & Historical Society.)

The vertical and incline hoisting works are apparent in the image of the Mammoth Mine, formerly known as the Old Boot Mine, in the Silver Bell Mountains. Mining in the Silver Bell Mining District began with multiple mining claims developed in the 1860s. The Old Boot Mine, established in 1865, was one of the earliest mines in the area. Mining was intermittent until the early 20th century, when the Silver Bell Mining Company was taken over by the Imperial Copper Company and later by ASARCO. (Courtesy of Albertype Company.)

This 1912 photograph depicts another early view of the Silver Bell Mining District, including the 450-foot union vertical shaft and powerhouses. It produced oxide and sulfide ores. (Courtesy of Albertype Company.)

Mining activity continued around the town of Silverbell until 1930, with a peak in population of 1,200 in 1920. The town bustled with grocery stores, restaurants, a fire station, a movie theater, and less respectable activities, such as prostitution. The elevation around the town ranged from 2,500 to 4,000 feet. (Courtesy of Albertype Company.)

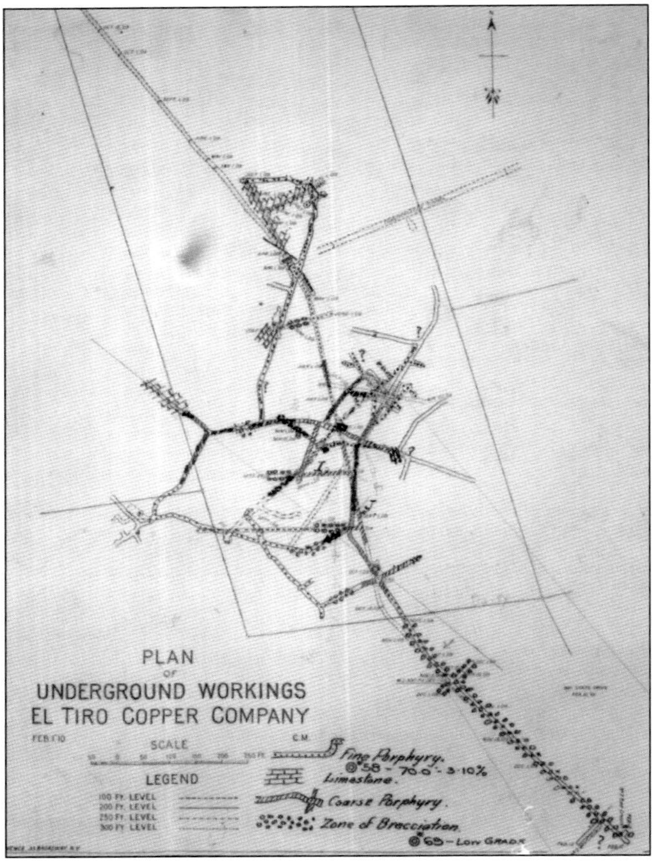

A blueprint shows the underground workings of the El Tiro Mine. (Courtesy of Albertype Company.)

Here is Silverbell as it appeared looking southeast in the early 1900s. The Imperial Copper Company established the town of Silverbell in the early 1900s over the remains of the original 1880s mining encampment. It remained the dominant company during the first decade of the 20th century, until its holds were acquired by ASARCO in 1919. In 1952, ASARCO resumed mining operations in the Silver Bell Mining District with an emphasis on open-pit development. In 1954, a new townsite established four miles to the southeast of the former townsite was also called Silver Bell. (Courtesy of Albertype Company.)

This early-20th-century image depicts the 350-foot-deep Billy shaft, known for its supply of lean cupriferous pyrite. (Courtesy of Albertype Company.)

Several locals walk the tracks at the mill in Silverbell. Located 40 miles northwest of Tucson, the Silver Bell Mine began as a series of mining claims and small mines under short-lived partnerships, companies, and lessees that dated back to 1865. The abundant oxide copper ores containing minor silver-lead ores brought about increased development by the turn of the century. (Courtesy of Albertype Company.)

This hoist and power plant operated at the El Tiro Mine in Silverbell during the first decade of the 20th century. Much of the ore mined at the El Tiro shaft, including chalcocite, cuprite, and native copper, was embedded in alaskite matrix. (Courtesy of Albertype Company.)

Here is a complete head frame at the Kurtz shaft with busy miners moving ore carts. Having started in the late 1800s, underground mining ceased at this location in 1931. (Courtesy of Albertype Company.)

This view, shown here in 2009, of the rugged topography of the Silver Bell Mining District looks at the northern portion of the Silver Bell Mountains and the El Tiro Pit directly below. (Courtesy of the author.)

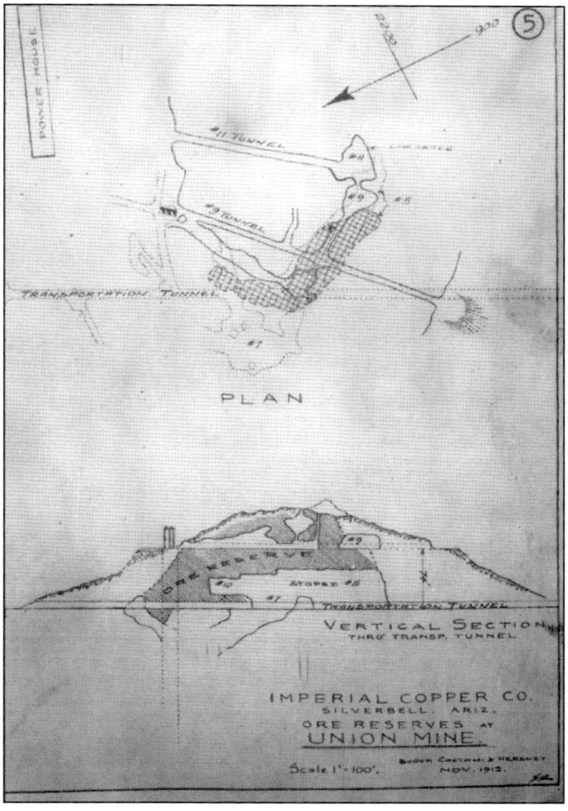

Depicting the deposits of the Union Mine, this blueprint features the underground traverse. (Courtesy of Albertype Company.)

In 1878, Ed Schieffelin discovered ore around Tombstone that led to the discovery of a silver ore bonanza. A series of successful claims followed, including the Tombstone, Graveyard, Lucky Cuss, and Toughnut. The discovery of silver in Tombstone convinced the Southern Pacific Railroad to expand eastward through Tucson instead of following a more northerly route along the Gila River basin into New Mexico. By 1882, Tombstone contained a population over 10,000. It was the largest town between San Antonio and San Francisco. Contention City, located on the San Pedro River, had three stamp mills that reduced silver ore from the Tombstone Mines. About 20 years later, the population dropped to a mere 1,000 because flooding hindered mining operations and the value of silver declined. Phelps Dodge acquired some of the mines in 1914; however, by 1950, the town became more of a tourist site than a mining town. Early mining ventures in Tombstone yielded over $38 million in metals. (Courtesy of Library of Congress Prints and Photographs Division.)

Four

CANANEA

Although not located in the geographical confines of southeastern Arizona, the town of Cananea and the Cananea Mine were influenced by their northern neighbor both economically and politically. The city of Cananea was established in 1901 and is located northeast of the 1.2-mile-diameter open-pit mine. The name is derived from the Apache Indian language and means "horse meat."

William Cornell Greene, famous for developing the rich copper porphyry deposits at Cananea, arrived in Arizona from the east in 1877. He owned a ranch in the San Pedro Valley along with multiple mining claims in southeastern Arizona, including Tombstone. In 1898, Greene acquired the Cobre Grande mining property in Cananea, valued at $47,000. Greene had the gift of gab and, since he only had several hundred dollars for investment, he convinced investors to take risks by buying his company stock. Relying on several partners from Tombstone to provide funding, Greene sensationalized the value of his first mine in Mexico, the Cobre Grande, at $250,000. The following year, the Cananea Consolidated Copper Company was established and became known as the "Four C's."

The birth of the Mexican Revolution of 1910 is often attributed by historians to labor disputes in Cananea in 1906 and the political unrest among many Mexicans who were unhappy under the iron rule of their president, Porfirio Díaz. During the time of the strike, Cananea had over 23,000 inhabitants. Mexican workers grew to resent the disparity of their wages and working conditions compared to their Anglo counterparts and decided to strike on June 1, 1906.

Colonel Greene needed to restore order. He called upon the Arizona Rangers. Upon reaching Cananea, the Bisbee volunteers hastened from their train to secure the smelter and concentrators. An enthusiastic crowd awaited them, including Colonel Greene. Reprimanding the crowd of workers for their acts of violence was Gov. Rafael Izábal, who was first to speak. Izábal was followed by Colonel Greene, who defended the company's wage preferences and the need to hire skilled American workers at competitive wages. Once the speeches were over, the Bisbee volunteers remained on the scene for the remainder of the day. Not a shot was fired. That afternoon, 75 *rurales* (mounted police) rode in under the command of the Russian-Mexican cavalry commander Lt. Col. Emilio Kosterlitzky. They were sent by President Díaz to quell the strike. Martial law was declared, and the violence ceased. Mexican casualties (over 20) outnumbered their American counterparts (6) at the end of the strike. A curfew was issued at the end of the day declaring that all persons, regardless of nationality, were to remain inside their homes or be shot on site if found on the streets. The immediate aftermath of the strike was a resumption of work at the mine. However, the lasting effects of the strike were far-reaching with Mexicans who grew to resent the invasion of a foreign power on Mexican soil. Four years later, the country was immersed in the Mexican Revolution.

During the 1890s, Cananea became a magnet for American investors. Mexican mining operations were closely watched by American entrepreneurs who sought to develop the land. They had the support of Mexican president Porfirio Díaz to develop Mexico's resources. The Mexican government was supportive of the mining industry. Mineral rights were deemed separate from surface rights, and claims could be kept indefinitely as long as their annual tax was paid. William Greene saw Cananea, Mexico, as a potential copper bonanza that could rival the mines at Butte, Montana. He convinced investors of the potential of the Cananea mining venture. Greene speculated heavily on his own stocks and used his capital to enhance his investment, such as convincing the Southern Pacific Railroad to build a spur line to Cananea in 1901 and building seven blast furnaces to treat 1,000 tons of ore a day the following year. Greene's spending habits were overindulgent, including acquiring a private railroad car and lavish living expenses in New York. Shortly after the strike in 1906, Greene was overcome by mounting debts and the loss of confidence among his stockholders. With his line of credit cut, combined with the impact of the labor strike and the merger with the Amalgamated Copper interests controlled by Standard Oil, Greene lost control of his company. The Financial Panic of 1907 further eroded Greene's fortune, forcing him to retire to his ranch. William Cornell Greene, the "Titan of the Southwest," whose reign of power lasted from 1899 to 1906, died August 5, 1911, having succumbed to pneumonia after having been thrown from a carriage. (Courtesy of Southern Methodist University, Central University Libraries, DeGolyer Library.)

Southeastern Arizona miners had long had an interest in the mines down in Mexico. This image dates back to 1883, the same year that William Greene was prospecting around Tombstone with aspirations to make a fortune south of the border. Mining interests in southeastern Arizona were hungry for new prospects in Mexico, and the Mexican government recognized the added benefit of the American dollar in its national economy. (Courtesy of the National Archives and Records Administration.)

Employees pose for a photograph at the site mine building. Sporadic mining operations in the area date back to 1760, when the Jesuits mined gold and silver under the crown of Spain. Gen. Ignacio Pesqueira, whose wife, Elena Pesqueira, later sold her claims, also mined the area in the late 19th century. During the 1880s, southeastern Arizona mining interests began to recognize the value of the mines in north Sonora, Mexico, most notably the Cananea porphyry copper deposit. (Courtesy of Southern Methodist University, Central University Libraries, DeGolyer Library.)

Employees stand outside the company office in Cananea, Mexico. From 1904 to 1906, Greene's company was at its zenith, controlling the largest copper ore body in Mexico. Profits were being made comparable to the Phelps Dodge operation in Bisbee, eight smelter furnaces operated, and 5,500 men were on the payroll. Greene also received a concession in the form of tax exemptions and 900 square kilometers of land to exploit from the Mexican government in exchange for maintaining an annual copper production above 250,000 tons a year. (Courtesy of Southern Methodist University, Central University Libraries, DeGolyer Library.)

A Cananea Mine employee stands by a rail engine nearby the mill. The track is evident in the background, and seen at the far right is part of the town of Cananea. (Courtesy of Cosme Rodriguez Sesma.)

Porfirio Díaz, seen here on his high horse, ruled Mexico with an iron fist from 1876 to 1911. His philosophy toward developing Mexico's resources was through foreign investment. William Greene and his business ventures, in developing the Cananea mine, received the full support of the Díaz government both monetarily in funding and militarily in suppressing the 1906 strike. (Courtesy of Library of Congress Prints and Photographs Division.)

Mexican workers are seen in the lumberyard during the strike of 1906. On June 1, 1906, events occurred involving the strikers and George Metcalf, who was both the manager of the lumber business and in charge of housing for Greene's Cananea Consolidated Copper Company, that would forever alter the relationship between labor and management in Cananea. The strikers proceeded that day to visit every unit of the Cananea Consolidated Copper Company to ensure that all work ceased. When they arrived at the lumberyard, George Metcalf refused them entry. They attempted to pass when Metcalf hosed them with water. Angered, they again tried to pass when gunfire broke out, killing three of the strikers. During the ensuing confrontation, both George Metcalf and his brother William were killed with miners' candlesticks. A fire started that consumed the lumberyard at a cost of $100,000. However, the violence was far from over. (Courtesy of Southern Methodist University, Central University Libraries, DeGolyer Library.)

Colonel Greene, seen here addressing strikers in Cananea from his car, appears to have drawn a crowd on this day June 3, 1906. Governor Izábal sits behind Greene to his left listening intently to the speech. Greene's main concerns were simple: production and profit. (Courtesy of Southern Methodist University, Central University Libraries, DeGolyer Library.)

Colonel Greene gives a passionate address to the Mexican copper mine workers during the 1906 labor strike. The police department sign is clearly identified at the far back. By 1900, Greene became one of the wealthiest businessmen in North America because of his share in the mines in Cananea and his ability to acquire the capital and organization to run them. Greene formed a myriad of additional business ventures established to help his mining enterprise, which consisted of cattle, coal, timber, and railroads. His large holdings in cattle included the Cananea Cattle Company in Mexico and the Greene Cattle Company in Arizona. He also started the Greene Gold-Silver Company and the Greene Consolidated Gold Company. (Courtesy of Southern Methodist University, Central University Libraries, DeGolyer Library.)

The Arizona Rangers were a tough breed of lawmen modeled after the Texas Rangers. The original Arizona Rangers' period of existence was brief, spanning the years of 1901 through 1909. However, during that time, they managed to bring law to southeastern Arizona with the barrel of a gun. They certainly made an impression on the striking Mexican workers in Cananea, who came to resent the United States' involvement on their sovereign soil. (Courtesy of Southern Methodist University, Central University Libraries, DeGolyer Library.)

A group of armed American Company employees and Arizona Rangers stands vigilant guard around the company store during the Cananea strike of 1906. Capt. Tom Rynning of the Arizona Rangers was in Bisbee on the day of the strike. He received a call from Colonel Greene expressing the dire predicament in Cananea and how every American's life was in danger. Excitement prevailed in the streets of Bisbee, as several thousand people gathered in the plaza awaiting word from Cananea. About 275 volunteers gathered in front of the Copper Queen Hotel under the command of Arizona Ranger captain Tom Rynning. Against the warning of Arizona territorial governor Joseph Henry Kibbey, they proceeded to cross the border by rail as civilians at Naco, where the Bisbee volunteers were met by Governor Izábal and inducted into the Mexican army. This was done because the Mexican Constitution would have deemed an armed force crossing the border an invasion. (Courtesy of Southern Methodist University, Central University Libraries, DeGolyer Library.)

David Olivarria examines the workings of the power plant in Cananea, ensuring its performance passes inspection. (Courtesy of Cosme Rodriguez Sesma.)

Benito Sandovel, an engineer for Compania Minera de Cananea, has just returned from an ore shipment. (Courtesy of Cosme Rodriguez Sesma.)

Seen here working on the railroad at Cananea are Mexican miners, who accused the American mine owners of giving American workers higher wage preference. Their average pay in 1906 was three pesos a day, compared to over five pesos a day for their American counterparts. Their requests included a raise in pay to five pesos a day along with a reduction in daily work hours from 10 to 8. They also wanted Mexican foremen to oversee their work instead of the standard practice of employing American foremen. (Courtesy of Cosme Rodriguez Sesma.)

This photograph depicts a typical day at the mine for Mexican workers between shifts outside the Oficina del Tajo during the 1960s. In 1907, William Greene was removed from the presidency and from the board of directors. The remainder of a large portion of his fortune was lost in the financial panic that same year. Greene would live out the remainder of his days in seclusion on his ranch. He succumbed to death as a result of pneumonia instigated by injuries incurred by a carriage accident in 1911. The Anaconda Copper Company took over control of the mines in Cananea until 1971, when the Mexican government and private investors acquired the majority of the stock. By 1982, the Mexican government bought out Anaconda's remaining shares and then nationalized the Cananea Mining Company. In 1990, Grupo Mexico became the major stockholder of the mines, assuming control of the operation to the present. (Both, courtesy of Cosme Rodriguez Sesma.)

This aerial view of the Cananea Consolidated Copper Company depicts the smelter in operation. Located 25 miles south of the international boundary between southeastern Arizona and Sonora, Mexico, the Cananea Mine is one of the world's foremost producers of copper and has historically been tied to southeastern Arizona through American capital, labor, and technology. It is one of the earliest open-pit mining operations in North America and the largest in Mexico, producing over 40 percent of its national output. It produced over 164,000 tons of copper in 2006 before a worker strike over working conditions and health care halted mining operations the following year. As of September 2010, the Cananea mine went back into production after a three-year labor strike. (Courtesy of Cosme Rodriguez Sesma.)

This 1960s aerial view of downtown Cananea depicts the street A.V. Juarez and the avenue Col. Los Pinos Escuela Secundaria, which honors the Mexican miners who were killed during the 1906 labor strike. The smelter stack is evident in the distance. (Courtesy of Cosme Rodriguez Sesma.)

A photograph of the Cananea Copper Mine in Sonora, Mexico, was taken in outer space by the crew of Expedition 16 to the International Space Station on March 3, 2008. The Colorada Pit, stretching over a mile in diameter, is clearly evident at the upper right with its concentric benches surrounding its periphery. The reservoir to the lower left is a leach pool of water pumped from the mine workings. (Courtesy of NASA.)

Five

The Process

The Basin and Range Province, which covers southeastern Arizona, is composed of broad, low-elevation valleys separated by a series of parallel mountain ranges that extend from northern Mexico to Idaho. The landscape was shaped over 25 million years ago by volcanism, forming both mountains and valleys throughout the region. Over millions of years, these mountains have eroded several thousand feet to current elevations between 3,000 and 10,000 feet across southeastern Arizona.

Plate tectonics and volcanism, along with the natural process of weathering from water and wind, established the necessary ingredients for mineral formation in southeastern Arizona. Metals and metallic mineral deposits occur where movement of the earth's plates allows hot fluids to chemically and mineralogically modify existing rocks. The fluids and heat cause elements to combine, forming minerals. Over time, weathering from water and wind at or near the earth's surface further affects the rocks. The combination of these natural processes formed many of Arizona's copper deposits. Because the copper deposits in Arizona often contain large and small crystals together, they are known as porphyry copper deposits. These deposits are most notable south of the transition zone, below the Colorado Plateau in the Basin and Range region. During the past several hundred years, miners developed many processes to efficiently extract the maximum mineral content at an economic price in the region.

Copper mining involves a process that includes exploration, mine development, production, and long-term reclamation. Today, exploration involves the use of the global positioning satellite system and sophisticated mapping equipment. Early mining exploration included techniques such as attention to topography. This involved surface association by sight of volcanic areas most notably associated with faults, folds, and fractures. Mineralized deposits may be present near the exposed outcrop intersection of veins. An outcrop of quartz is a good indicator of gold and silver deposits. Drilling for core samples affirms the exploration of potential deposits to a depth of over 100 feet. During copper production, sulfide ores are moved to a concentrating stage involving crushing, fine grinding, and concentration by flotation, while oxide ores are routed to leaching tanks and mixed with acid with the dissolved copper recovered by precipitation on scrap iron or by the electrowinning process. Grinding mills and smelters, along with SX/EW (solvent extraction/electrowinning), process millions of tons of copper ore in Arizona every year. Copper is essential to the mass market, having uses in electrical equipment, plumbing, motors, power generators, transformers, lighting, coins, and many other needful machinery.

The 19th-century prospector, an opportunist by nature and a transient by profession, led a challenging life. There was always the need to discover another prospect in the hope of finding a mineral bonanza. Sometimes the prospect was worked, and other times it was sold for a modest sum to a prospective buyer if it proved a decent yield of mineral product. Many nights the prospector slept under the stars, while his diet consisted of tobacco, black coffee, bacon, and bread. The burros seen here standing by the galvanized iron tubs in the town of Ruby were the pack animals of choice for the western prospector. They can carry loads up to 400 pounds with the ability to travel up to 25 miles a day. They are also resilient toward extremes, including the heat of the desert and the cold of high altitude. (Courtesy of Tallia Pfrimmer Cahoon.)

A prospector is seen here panning for gold. The average size of the pan is between 9 and 16 inches in diameter. The process involves submerging the pan underwater and then discarding the larger material that accumulates. The pan is shaken, leaving only the gold and black sand at the bottom, a common ingredient in alluvial and placer deposits. (Courtesy of Arizona Board of Regents for the University of Arizona.)

It was the lot of the mule or burro to work the *arrastra*. Jose Cuestra's arrastra and burro, along with Arizona oak trees in the background, are seen in this photograph taken by Walter S. Pfrimmer in 1930. The arrastra ("to drag") was used both by the Spaniards and Americans to crush ore. The arrastra had two advantages: it was cheap to erect and it used animal power. It consisted of promenading horses or mules around a circular floor that was fashioned out of andesite blocks compacted with clay. A center wooden post was coupled with several horizontal beams and 200-pound dragstones. It was these implements, a trickle of water, and crushed ore that aided in the process of separating the metal from its parent rock. (Courtesy of Tallia Pfrimmer Cahoon.)

Later, mercury was used as a catalyst to capture gold and silver, a process called amalgamation. Depicted in this image is a homemade retort made from iron pipe fittings. Heated mercury and water is vaporized and condensed in this apparatus, with the solid material, such as gold or other precious metal, left behind in the retort. (Courtesy of Arizona Geological Survey.)

Gold placers are accumulations of gravel and sand that contain concentrations of gold eroded by natural weathering processes from gold-bearing rock. Eventually, deposits are carried downstream. A sluice is depicted in this photograph with a miner actively panning for gold among the jumbled rocks. Gold is six times heavier than ordinary rock and often settles at the lower gradient of a stream. The sluice is a sloping trough used to capture gold from placer sand and gravel. Running water is used to wash the lighter sand and gravel down and out of the sluice, while the heavier gold is captured behind the riffles. (Courtesy of the Arizona Geological Survey.)

Known among the miners as Bullet, this workhorse is seen pulling an underground ore cart in the No. 1 shaft of the Consolidated Tunnel in Dos Cabezas. Horses were often used for tramming trains of ore cars from the stopes. They hauled the ore to the shaft, where it was then hoisted to the surface. (Courtesy of SPV Arts & Historical Society.)

Carrying supplies in the Silver Bell Mining District is a 12-car compressed-air locomotive. In 1904, the Arizona Southern Railroad, a subsidiary of the Imperial Copper Company, began operating between Red Rock and Silverbell. (Courtesy of Albertype Company.)

The role of the compressed-air locomotive, shown above, included hauling ore and supplying air to the air drills in the mine. These locomotives are also said to have provided the mine with ventilation. This air engine provided service to the Imperial Mines in Silverbell during the early 20th century. The image below depicts six cars of ore hauled by a compressed-air locomotive. (Both, courtesy of Albertype Company.)

Fred Clifford stands in the foreground of the Central Copper Company assay office in Dos Cabezas. The purpose of the assay office was to test the purity of metals. (Courtesy of SPV Arts & Historical Society.)

The blacksmith shop served as an essential part of the mining town, manufacturing and maintaining milling equipment, ore trucks, and mining implements, such as picks, wedges, and hand steel for drilling. Frank Snyder, seen in this photograph, served as the chief blacksmith in Dos Cabezas. (Courtesy of SPV Arts & Historical Society.)

Two men are working on a steam boiler at Ruby in 1930. Walter Strange Pfrimmer is in the far back smoking a cigarette. (Courtesy of Tallia Pfrimmer Cahoon.)

Surveying instruments are evident in this photograph taken at Ruby, including both an alidade and plane table. The miner at left is using the alidade, a rule equipped with telescopic sights, to measure distance. Walter Strange Pfrimmer is seen at the far right. A Brunton compass, which measures angles and directions for mapping, is clearly seen mounted on a tripod at the far right. A miner's lunch box is also seen in the photograph, which was taken by light of the carbide lamp held by the third man from right. (Courtesy of Tallia Pfrimmer Cahoon.)

A miner is seen pushing an ore cart on the mine pad at the Montana Mine in Ruby. The ore cart has a 1,600-pound capacity. The elevator is clearly seen at left, and a galvanized iron roof is evident in the background. The two main metals mined at Ruby were lead and zinc. The lead went to the smelter in El Paso by rail, and the zinc went to Amarillo, Texas. (Courtesy of Tallia Pfrimmer Cahoon.)

Two miners are depicted in this photograph comprising a double-jack team. One is turning the steel, while the other pounds away in cadence. The eight-pound double jack the miner is holding is clearly making an impact on the steel rod chisel-bit tip in his partner's hand. Throughout the history of mining, miners have always been considered a tough lot. Miners in Arizona were no different. Prior to mechanization, miners relied on their endurance and fortitude along with their ability to swing a hammer or turn a steel spike. (Courtesy of Library of Congress Prints and Photographs Division.)

The illuminated shaft is provided courtesy of a single candle attached to an iron rod candle-holder embedded in a timber. A pick appears to add additional support or prop to the candle. The miner in the foreground with the chisel in hand and his compatriot, barely visible to the right holding the hammer, appear to be making progress at the face of the tunnel. (Courtesy of the National Archives and Records Administration.)

This group of miners is standing by a hoist or elevator, which was commonly referred to as the cage. It is rumored that miners disliked this contraption because of its lack of sides and overhead enclosure. Later improvements to the cage added wire mesh to remedy this oversight and a safety brake to prevent unscheduled drops. The elevator depicted here appears to have the later improvements. Third and fourth from left are mining engineers Walter S. Pfrimmer (in white pants) and Frank Lerchen (leaning against a pillar) in the Montana mine at Ruby around 1930. (Courtesy of Tallia Pfrimmer Cahoon.)

Men at work in a mining shaft house in Cananea pose for the picture above in the early 1900s. The heavy equipment in the background is a steam-driven hoist. The operator received a higher wage than his counterpart, who worked down in the shaft. The hoist operator shown below worked for the Central Copper Company in Dos Cabezas, Arizona. (Above, courtesy of Southern Methodist University, Central University Libraries, DeGolyer Library; below, courtesy of SPV Arts & Historical Society.)

A Dos Cabezas miner poses for the camera with his carbide lamp evident on the floor of a stope. Carbide lamps date back to the 1890s, when acetylene, a gas that burns with a bright, yellow-white flame, was developed. Carbide, although inexpensive, can be explosive when mixed with water and oxygen. A carbide lamp burns acetylene gas generated by the chemical reaction of water on calcium carbide. Miners have been killed by these explosive effects, and many fires have started underground because of acetylene. (Courtesy of SPV Arts & Historical Society.)

Miners are seen operating pneumatic drills in the Collins shaft at Tiger in 1940. This mining implement replaced the tedious process of hand drilling; however, it soon became known as the "widow maker" because of the dust it produced in the air. The dust caused silicosis among miners who inhaled the particles, which damaged their lungs, leading to scar tissue build up and death. Eventually, water was applied to keep the dust level down, adding years to the miners' lives and also making the pneumatic drill more efficient. (Courtesy of Magma Copper Company.)

A charge is being laid at the Cananea Mine. The holes are loaded with a mixture of ammonium nitrate and fuel oil (ANFO). (Courtesy of Cosme Rodriguez Sesma.)

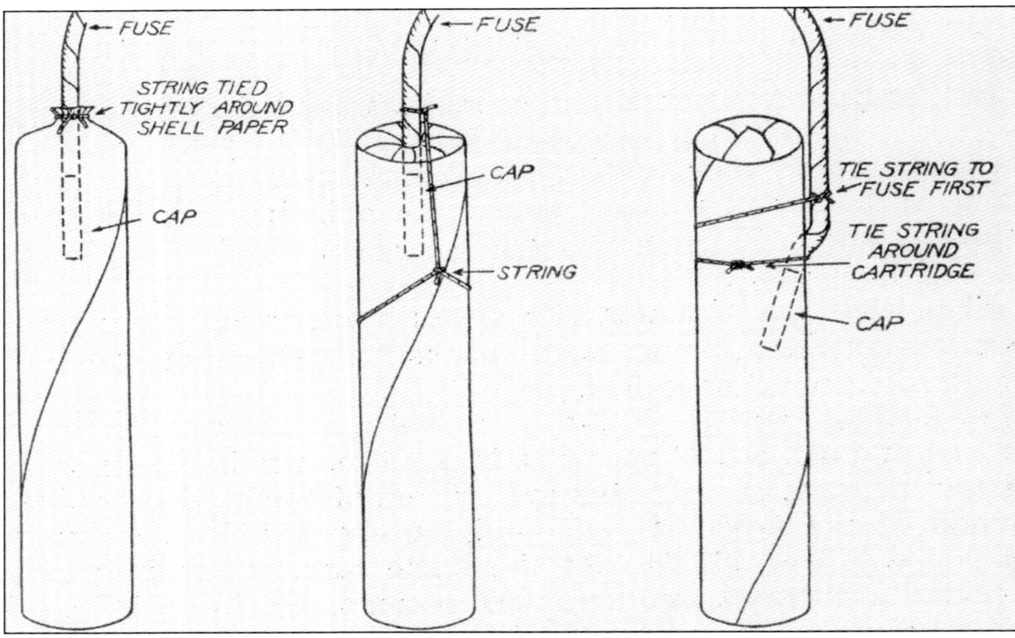

Prior to dynamite, miners relied on the explosive effects of black powder. In 1847, the Italian chemist Ascanio Sobrero invented nitroglycerin. Several decades later, Alfred Nobel invented dynamite. Dynamite sticks expanded the marketable use of nitroglycerin. They consisted of nitroglycerin mixed with a soil called Kieselghur and a sawdust filler along with ammonia nitrate. They could be retrofitted with the necessary charge to accommodate hard rock blasting. Proven safer to handle, dynamite replaced the volatile raw nitroglycerin, which had incited casualties among its users as a result of its explosive volatility. Sold in wooden cases averaging 50 pounds, cartridges of dynamite wrapped in tough paper became volatile at temperatures exceeding 350 degrees. (Courtesy of Arizona Board of Regents for the University of Arizona.)

Depicted in the photograph above, taken at the Ray Mine, the effect of a charged blast of explosives is reducing the rock to manageable pieces. Below is a shovel emptying low-grade copper ore into a 320-ton truck to be hauled off to the mill. It takes four scoops of rock to fill a truck of this size. The tires alone measure 11 feet in diameter. Clearly seen are the 40-foot-high benches commonly seen in open-pit mines. Today, copper ore bodies averaging two-percent copper can be extracted by the use of modern technology. The average mine produces only about 10 pounds of copper per ton of ore. For each ton of ore, an additional two tons of barren rock, or overburden, must be moved. Copper was one of the better performing metals on the market in 2010, with a price averaging $3.50 a pound. (Both, courtesy of the author.)

A typical early-20th-century 10-stamp mill appears in this image. It was powered by either steam or water to pound ore into fine powder by the use of its rising and falling iron stamps. (Courtesy of Arizona Board of Regents for the University of Arizona.)

It is the job of the crusher seen below to reduce the blasted pieces of rock to an average eight inches in diameter. (Courtesy of Cosme Rodriguez Sesma.)

Afterwards, conveyors feed the ore to massive horizontal cylinders called mills, where it is further reduced or ground to a powder containing less than one percent of copper by the friction of steel balls, long steel rods, or sometimes just the ore itself mixed with water. (Courtesy of Cosme Rodriguez Sesma.)

After the sulfide ore is crushed at the mill, it is sent to the concentrator, where the waste rock and slag are removed. Ore concentration by flotation is depicted in this photograph of a "middlings thickener," a 400-foot diameter tank located in Cananea, Mexico. Slurry is mixed with reagents, such as ferric sulfate and nitric acid, along with ground limestone, pine oil, alcohol, and oxygen to extract the copper. The denser solid waste material sinks to the bottom and is emptied out of the tank. The valuable chalcopyrite grains react with the mixture and rise to the surface, flowing over the edge of the tank. Ore valued at 28-percent copper is obtained in this process. The water in the concentrator is recycled for further use, and the waste slag is removed. (Courtesy of Cosme Rodriguez Sesma.)

This photograph of the Cananea mine depicts the appearance of the ore after it is milled and concentrated. Called slurry, it bears resemblance to a thinly diluted watery mud that is 15-percent copper. (Courtesy of Cosme Rodriguez Sesma.)

Oxide ore found in tailings is sent through a leaching process. The tailings are leached by a weak acid solution to recover the maximum copper, as shown in this image of a leach pond in Cananea. (Courtesy of Cosme Rodriguez Sesma.)

The open-pit mine in Cananea is pictured from above. (Courtesy of Cosme Rodriguez Sesma.)

This picture was taken looking west above the Sierrita-Esperanza Mines toward the Sierrita Mountains and distant Baboquivari Peak, located in the Baboquivari Mountain Range. The middlings thickners (several hundred-foot-diameter tanks in the foreground) denote ore concentration by flotation. Discovered in 1955, the Esperanza ore body, owned by the Duval Corporation, became an open-pit mine over the next decade with the addition of a 60,000-ton-per-day concentrator. Minerals mined at this locality include copper, molybdenum, and silver. (Courtesy of Alvin F. Lindstrom)

This photograph at the Cananea Mine depicts a flash-smelting furnace, whose job it is to separate the iron and sulfur from the copper. Ground material is placed into a furnace under extreme heat. An oxygen flash furnace heated with natural gas contains a silica flux, which binds with both the iron and other impurities, keeping them separate from the copper. Matte is poured into a ladle, while an overhead crane transports another ladle filled with molten matte to a converter furnace, where it is exposed to air to burn away both iron and sulfur. The resulting product is blister copper, molten metal 98-percent pure. (Courtesy of Cosme Rodriguez Sesma.)

The anode furnace, seen in operation, is blowing natural gas into the smelting process of metal to burn off the oxygen. Note the light emanating from the top and side of the anode furnace. It starts out a yellow flame, turning blue-green once the oxygen has been burned away. The purpose of the anode furnace is to remove the excess oxygen. The separated copper will next be cast into slabs. A single slab, known as an anode, can weigh as much as 750 pounds. (Courtesy of Cosme Rodriguez Sesma.)

Although these anodes contain 99-percent pure copper, they fail to meet the specifications of manufacturers because of impurities. The next step involves shipping the anodes to a refinery to remove the impurities. Mixed with copper sulfate along with sulfuric acid while charged with electricity, a purer form of copper at 99.99 percent is established as a cathode weighing nearly 300 pounds. (Courtesy of Cosme Rodriguez Sesma.)

Depicted in this photograph taken at an SX/EW plant in Morenci are high-quality copper cathodes, the resulting product at the solvent extraction/electrowinning process plant. This electrolytic process produces 99.9-percent pure copper, which is also known as electrolytic copper. (Courtesy of the author.)

Six

The Product

Mineral collecting, a hobby made more accessible by both past and present mining endeavors, is an addictive hobby for southeastern Arizona rockhounds. It takes access, research, and time to obtain these mineral treasures from deep in the earth in mining towns of southeastern Arizona. Accessibility to mine sites involves networking and the permission of the mine owner. Involvement with a mineral collecting club is always a benefit because of the shared knowledge base among colleagues and field trip opportunities. A geologist's rock hammer, hard hat, steel-toe boots, leather gloves, five-gallon bucket, tissue paper, and a crowbar are implements that aid the collector in mineral extraction and collection.

One should not discount the benefits of prior research of the destined mining town or locality. Publications such as Arizona Bureau of Mines bulletins and online sources like Mindat are invaluable for updated mine site information, including types of minerals, history, and accessibility. Arizona Geological Survey also provides the public with educational information on the geology of Arizona through maps and publications. Notable museums in southeastern Arizona with an emphasis on mining and mineral history include the Bisbee Mining & Mineral Museum, Flandrau Science Center, and Arizona-Sonora Desert Museum.

The Tucson Gem & Mineral Society is the premier mineral collecting club in southeastern Arizona and is renowned for sponsoring the internationally acclaimed two-week Tucson Gem & Mineral Show, held annually at the end of January. This is the largest, most diverse mineral show in the world, attracting thousands of visitors from around the world and across the country while generating over $100 million in revenue toward the local economy. Contributing to the success of the Tucson Gem & Mineral Show from its humble origins in 1955 to the present is its proximity to the mineral-rich mining towns and localities in southeastern Arizona and the support of the Tucson Gem & Mineral Society, along with the support among the mineral dealers and buyers both local and international.

The very first Tucson Gem & Mineral Show is a real eye-opener for these children, who appear to be fixated on the crystal display. From left to right are Richard Jones, Charles Jones, and Ann Fordham. First held at the Helen Keeling School's auditorium on March 19 and 20, 1955, the show garnered 1,500 people during its first year, which had the benefit of free admission. From that day forward, the show became an annual event in Tucson. (Courtesy of Tucson Gem & Mineral Society.)

Smithsonian curator Paul Desautels attended the Tucson Gem & Mineral Show in 1961. It was the first time the Smithsonian had participated in any club event. From that day forward, Paul Desautels maintained a strong presence at the show through notable lectures, publications, and mineral displays. (Courtesy of Dan Caudle.)

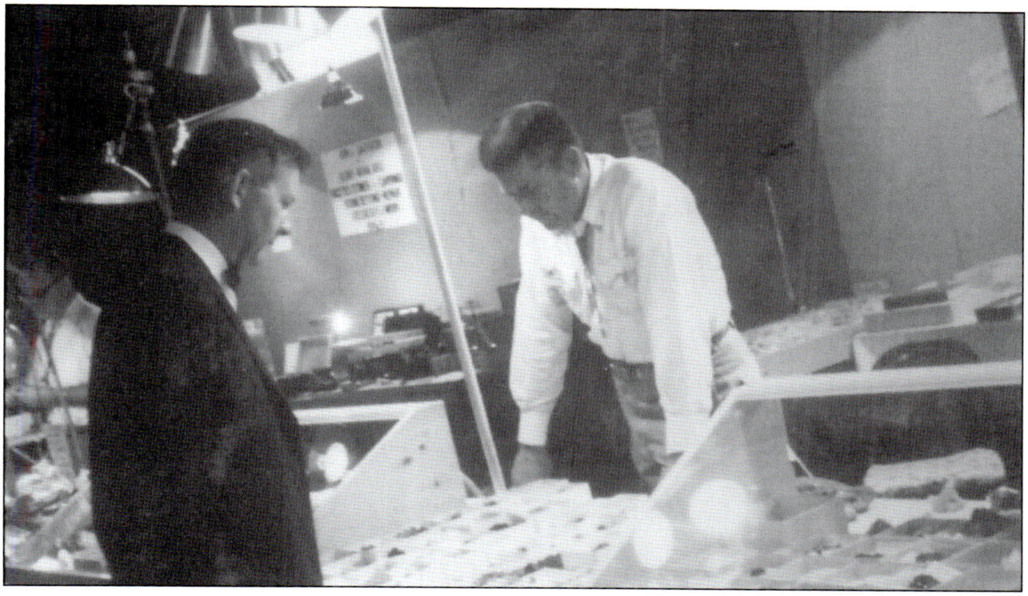

Dan Caudle presides over his mineral collection during the 1959 show. He acquired his interest in both gems and minerals during his service with the Army Air Corps in New Guinea during World War II. He held the distinction of being the first show chairman of the Tucson Gem & Mineral Show in 1955. The 1959 show included two organized collector field trips to the Inspiration Consolidated Copper Mine in Globe-Miami and the Pima Mine at Twin Buttes. (Both, courtesy of Tucson Gem & Mineral Society.)

Bob Roots, an enterprising mineral dealer from Denver, Colorado, successfully advocated the establishment of the first Tucson Gem & Mineral Show. (Courtesy of Tucson Gem & Mineral Society.)

Joseph Normart, an avid mineral collector and metallurgist, studies a mineral specimen. Arriving in Tucson at the age of 16, he was employed at his uncle's foundry shop off Congress Street. For many decades, he collected minerals from such localities as Tiger, St. Anthony, and Cerro, Colorado, and he displayed his minerals during the 1961 Tucson Show. (Courtesy of Tucson Gem & Mineral Society.)

Displayed at the seventh annual Tucson Gem & Mineral Show, this scale working model of the anode furnaces and casting wheel at Kennecott's Ray Mines Division certainly captivated the audience depicted in this photograph. John A. Morrow is seen operating the model, while Jim Maize is discussing the specifics of the operations. (Courtesy of Tucson Gem & Mineral Society.)

The mineral being held up, either galena or fluorite, prompts a discussion at the 1962 Tucson Gem & Mineral Show. Those pictured are chairman Bill Schupp at left and an unidentified gentleman at right. The Smithsonian participated in the event the previous year, marking its first participation in any club event in the nation. (Courtesy of Tucson Gem & Mineral Society.)

The 15th annual Tucson Gem & Mineral Show draws a crowd at the Pima County Fair and Rodeo Grounds in these 1969 photographs. The building was a World War II–era Quonset hut that hosted the show from 1956 until it was relocated to its present location at the Tucson Convention Center in 1972. One of the star attractions at this year's show was the brilliant yellow crystals and sprays of the mineral legrandite from Mina Ojuela, Mapimi, Durango, Mexico. (Both, courtesy of Tucson Gem & Mineral Society.)

These images represent a Tucson Gem & Mineral Show–sponsored field trip out to the Ray Mine. The minerals that are being collected this day include chrysocolla and native copper. The above image includes a distant view of Tea Pot Mountain. This renowned mineral collecting site hosts a diverse collection of minerals, including brochantite, cuprite, descloizite, dioptase, libethenite, and torbernite. (Both, courtesy of the author.)

This sample of aragonite ($CaCO_3$), measuring four and a half inches by two and a half inches, is from a limestone cavern in the older workings of the Copper Queen Mine in Bisbee, Arizona. It appears with a transparent to translucent glassy luster. Composed of colorless or white prismatic crystals, aragonite contains a subconchoidal fracture with a colorless streak. Noted for its coral-like habit and resinous luster, its occurrence is widespread; the mineral is found in limestone caves, mineral veins, and hot springs. Named in 1797 for the type locality, Molina de Aragón, Spain, aragonite possesses the quality of florescence, giving off a blue, pink, or yellow color when exposed to ultraviolet light. Over time, aragonite alters into calcite, such as in the case of fossil shells. Its uses range from collectible purposes to industrial ones like cement and steel production. (Courtesy of Kenny Don.)

This stunning example of brochantite ($Cu_4SO_4[OH]_6$), a hydrous copper sulphate from weathered zones of copper deposits in the Millpillas Mine in Cuitaca, Municipio de Cananea, Sonora, Mexico, possesses acircular, shiny, emerald-green crystals. Bronchantite was named after the 19th-century French mineralogist A.J.M. Brochant de Villiers. This specimen measures five and a half inches by four inches. Additional qualities of brochantite include a green, vitreous luster whose streak is pale green with an uneven-to-conchoidal fracture. (Courtesy of Kenny Don.)

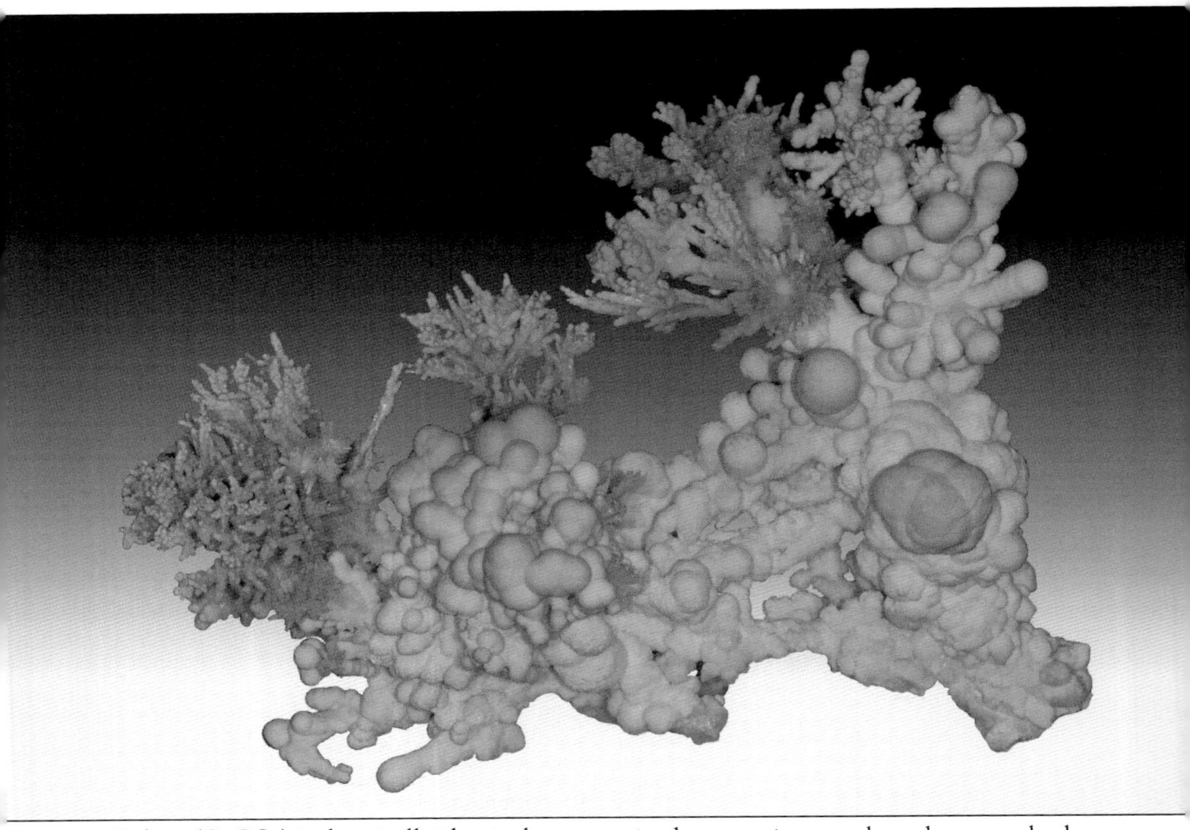

Calcite ($CaCO_3$) is chemically identical to aragonite; however, its crystals are hexagonal, whereas aragonite possesses an orthorhombic system with acircular crystals. One of the most common minerals on earth, calcite derives its name from *chalix*, the Greek word for lime. Found in caves, it forms stalactites and stalagmites precipitated from water. It also forms from the shells of marine animals. This five-inch-by-four-inch specimen hails from the Higgins Mine in Bisbee. Calcite's industrial uses include cement, glass, and steel. Some important varieties of calcite include marble, onyx marble, and travertine. Calcite has the quality of white-blue fluorescence under short-wave ultraviolet light. It also has the quality of white phosphorescence for an extended period of time after the short-wave ultraviolet light is removed. (Courtesy of Kenny Don.)

Cerussite ($PbCO_3$), a minor ore of lead, is a highly collectible mineral. Named for the Latin term *cerussa*, meaning "white lead," this nine-inch-by-six-inch specimen from the Flux Mine in the Patagonia Mountains of Arizona appears with vertically striated crystals and jackstraw clusters upon its rock matrix that are reminiscent of skeletal lattices or bridge girders. Cerussite possesses qualities of being transparent to translucent with apparent fibrous white crusts of loosely consolidated needles. Operated from the 1850s through 1963, the Flux Mine contributed over half of the ore tonnage mined in the Harshaw District. During the early 1860s, cerussite from the Flux Mine was rumored to have been one of the lead ores contributing to the manufacture of bullets for the Confederate army. (Courtesy of Kenny Don.)

A three-quarter-inch-by-two-inch specimen of copper (Cu) from San Manuel, Pinal County, Arizona, is depicted in this photograph. Arizona miners in the mid-19th century disregarded this metal as having little value. Copper was considered the ugly duckling of minerals until the latter decades of the 19th century, when electrical uses increased demand during the Industrial Revolution. Because copper is an excellent conductor of both electricity and heat, it has served as a source for industry during the last 150 years. Arizona accounts for over 60 percent of copper production in the United States, giving it the title of the "Copper State." (Courtesy of Kenny Don.)

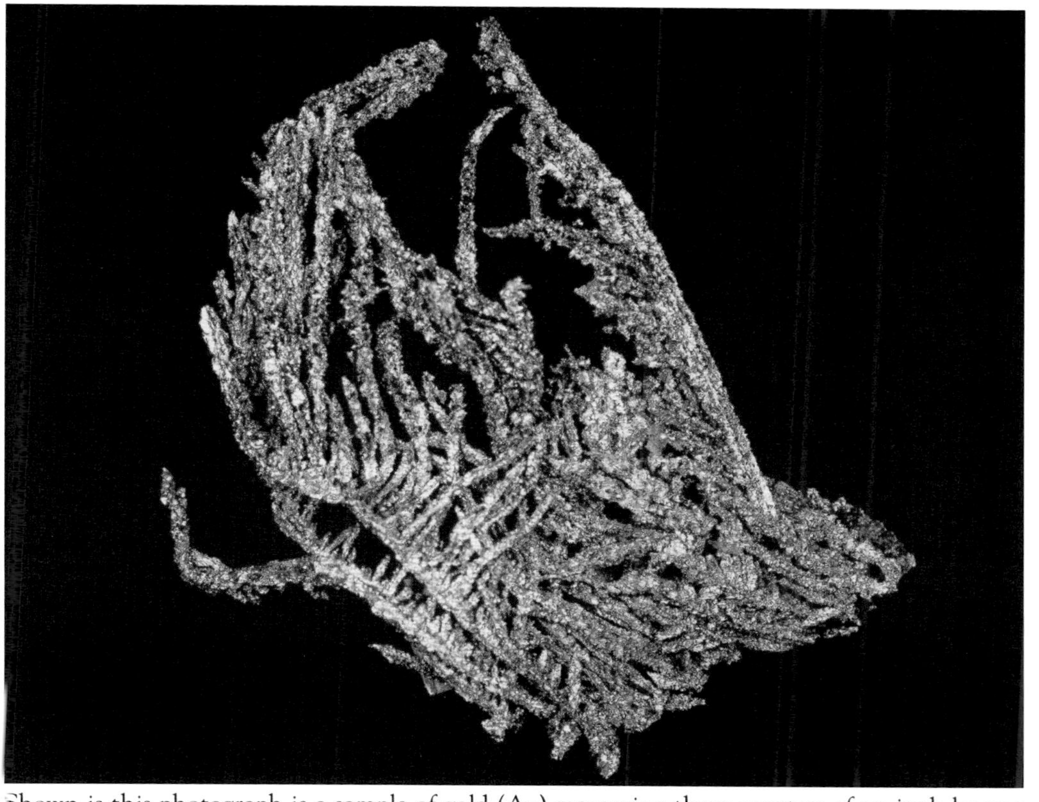

Shown is this photograph is a sample of gold (Au) measuring three quarters of an inch by one inch from the Greaterville District. Highly sought after by miners who discovered the rich placer deposits in the area during the 1870s, Greaterville placer gold is known for being bright and coarse and sometimes iron-stained. Flakes found at this locality average a 10th of an inch in diameter, while nuggets range from a quarter-ounce to the extremely rare find of 37 ounces. Gold has the properties of being the most malleable and the most ductile of all known metals. Like copper and silver, it is an excellent conductor of electricity and heat. It is also resistant to the natural effects of decomposition. Gold's scarcity on the earth's crust averages less than one seven-thousandth of an ounce per ton. Gold coin (seen below) was a lawful circulating medium of exchange in the United States from the beginning of the republic until it was replaced by fiat money in 1934. (Above, courtesy of Kenny Don; below, courtesy of the author.)

A pseudomorph is a mineral that takes the external appearance of another mineral species that it has replaced. The renowned Sistine Chapel, painted by 16th-century artist Michelangelo, provides an excellent example of mineral pseudomorphs. Painted with a vibrant blue paint derived from the powder of azurite, the sky mural turned from blue to green with the passage of several centuries. This was the result of water loss from the azurite pigment, which naturally altered the chemical makeup and crystal structure of azurite, transforming it into the green mineral malachite. This specimen of malachite pseudomorph after azurite from Bisbee, Arizona, measures two and three quarters inches by two and a half inches. (Courtesy of Kenny Don.)

This specimen of malachite ($Cu[CO_3][OH]_2$) from the famous Copper Queen Mine in Bisbee, Arizona, measures three and a half by three inches. The most common secondary copper mineral found in oxidized zones of copper deposits, malachite appears in brilliant green botryoidal encrusting masses and is named after the Greek *malache*, for "mallow herb leaf." A popular ornamental stone used in the manufacture of tabletops and vases, it is sold in a polished form at the Tucson Gem & Mineral Show as jewelry, paperweights, and trinket boxes. (Courtesy of Kenny Don.)

This one-and-three-quarter-by-two-and-a-half-inch amethyst variety of quartz (SiO_2) is from Washington Camp, located in the Patagonia Mountains in Santa Cruz County, Arizona. Amethyst is the purple variety of quartz commonly found in many types of rocks, including granites, gneisses, and mineral veins. Colored by impurities of manganese, amethyst is a popular ornamental stone. (Courtesy of Kenny Don.)

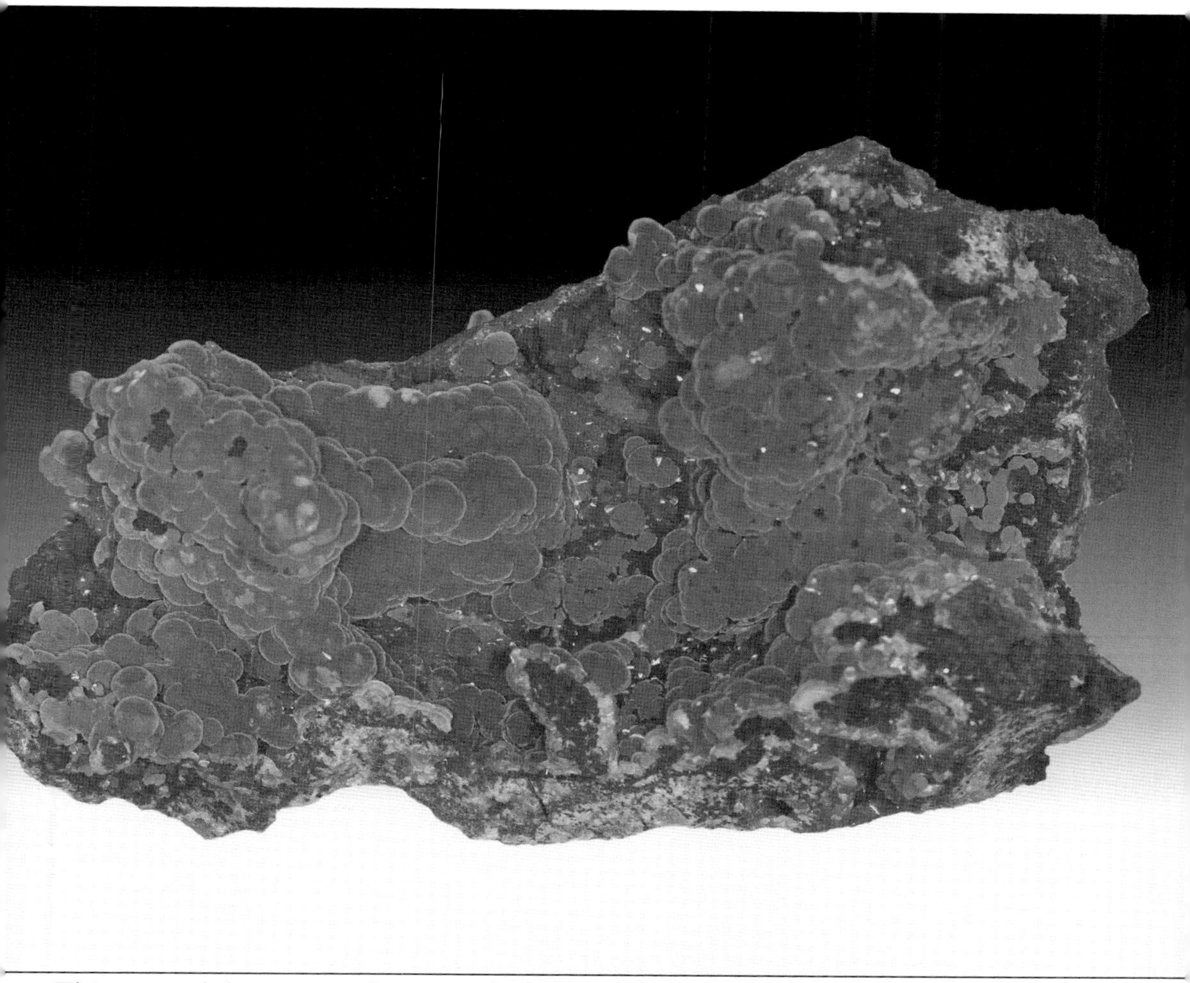

This one-and-three-quarter-by-two-and-a-half-inch botryoidal, teal-blue sample of rosasite ($[Cu,Zn]_2[CO_3][OH]_2$), a secondary mineral, was found in the oxidation zone of copper-zinc deposits from the Silver Bill Mine in Gleeson, Arizona. Named after its discovery locality, the Rosas Mine in Sulcis, Sardinia, Italy, in 1908, it is a popular piece to add to a cabinet specimen collection. Rosasite forms crust-like spherical, rounded aggregates on matrix. Its steak is light blue/green with a vitreous silky luster. Additional qualities of this copper zinc carbonate hydroxide include crystals which are transparent to translucent. Rosasite possesses the quality of effervescence in cold dilute hydrochloric acid. Over long periods of time, rosasite can fade and change color when exposed to excessive heat or cold. The fragile gem is seldom cut in cabochons and is often used as a decorative landscape rock. (Courtesy of Kenny Don.)

Discovered near the town of Superior, Arizona, in the legendary Silver King Mine, this sample of native silver (Ag) appears with curled silver wires, some of which exhibit dendritic patterns. Although renowned for its copper deposits, it was the search for gold and silver that initiated the mining industry in southeastern Arizona. Silver has the highest conductivity of all metals. It is the most ductile and malleable metal second to gold and is therefore used as silver plating and wiring for industrial purposes. Silver has traditionally been used in the manufacture of coins, electrical equipment, jewelry, photographs, and tableware. Silver coin (seen below) was a lawful circulating medium of exchange in the United States from the beginning of the republic until it was replaced by fiat money in 1965. (Above, courtesy of Kenny Don; below, courtesy of the author.)

This eight-inch-by-six-inch sample of vanadinite from the Old Yuma Mine includes a variety of hexagonal, prismatic crystals. Some specimens of vanadinite are highly sought after for their vivid red color. Vanadinite ($Pb_5[VO_4]_3Cl$) is the main ore of vanadium, an element that is renowned for its great industrial strength. For that reason, it is used in alloys, most notably of iron in the form of ferrovanadium. (Courtesy of Kenny Don.)

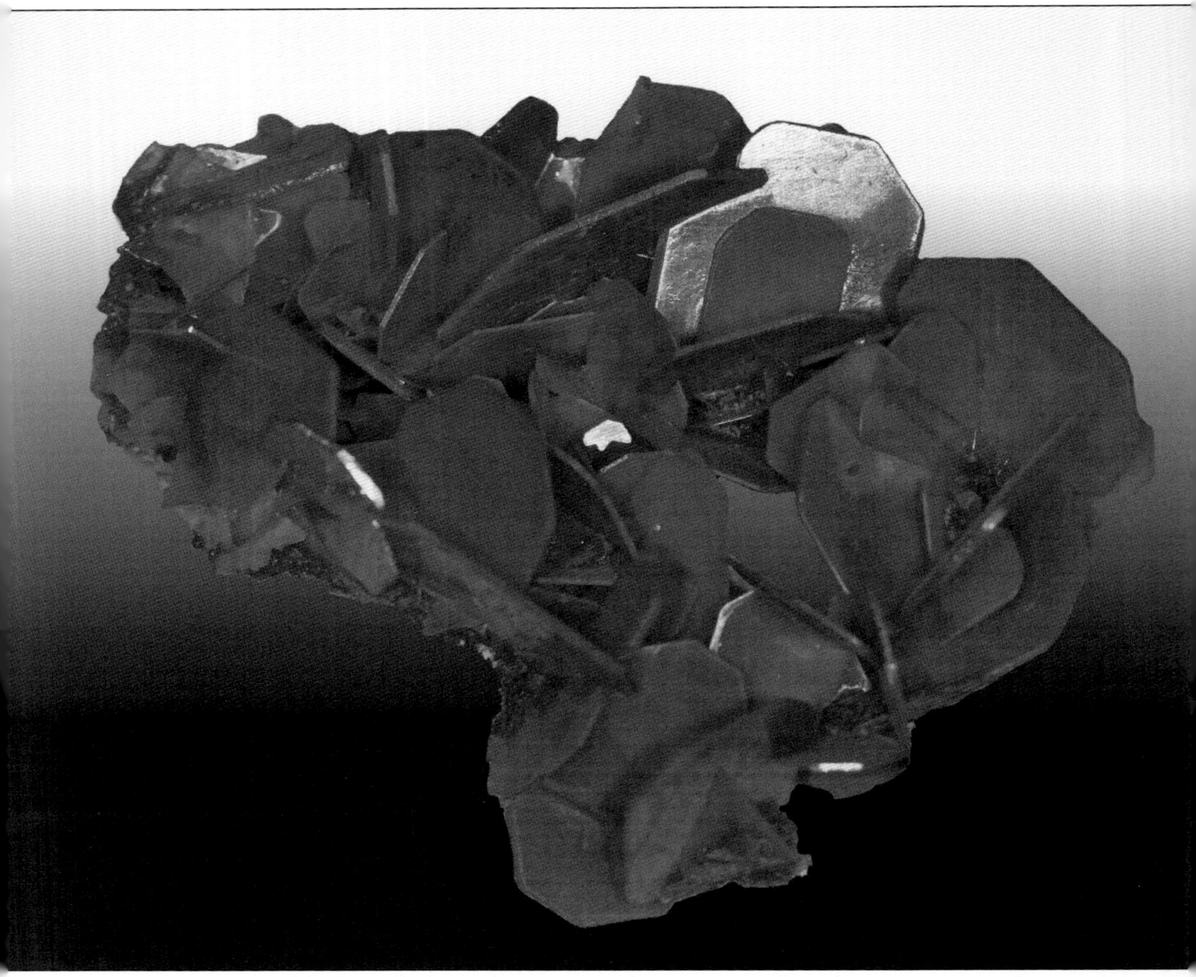

Lustrous yellow, blocky, tabular wulfenite crystals on gray oxide-rich matrix are evident in this seven-eighths-by-one-and-a-quarter-inch specimen from the Glove Mine in Santa Cruz County, Arizona. Named after Austrian mineralogist Franz Xavier Wulfen (1728–1805), wulfenite ($PbMoO_4$) crystals may exhibit the quality of twinning, defined as sharing similar crystal faces. Wulfenite is an ore of molybdenum whose uses include strengthening steel coupled with heat and corrosion resistance. Additional qualities of wulfenite include tabular crystals with a resinous to adamantine luster. The characteristics of fracture are subconchoidal to uneven. Wulfenite can be dissolved in concentrated hydrochloric acid. Some of the most spectacular wulfenite specimens have been collected at the Red Cloud Mine, La Paz County, and the Old Yuma Mine in Pima County. (Courtesy of Kenny Don.)

BIBLIOGRAPHY

"Arizona Lode Gold Mines and Gold Mining." *Bulletin No. 137.* Tucson: University of Arizona, Revised 1967.

"Arizona Zinc and Lead Deposits: Part II." *Bulletin No. 158.* Tucson: University of Arizona, 1951.

Bailey, Lynn Robinson. *Bisbee, Queen of the Copper Camps.* Tucson: Westernlore Press, 1983

Canty, J. Michael and Michael N. Greeley, eds. *History of Mining in Arizona.* vol. I–III. Tucson: Mining Club of the Southwest Foundation, 1987–1999.

"Exploration and Development of Small Mines." *Bulletin No. 164.* Tucson: University of Arizona, 1954.

Forrester, James Donald. *Principles of Field and Mining Geology.* New York: John Wiley & Sons, Inc., 1946.

Garza, Phyllis and Carol Wien. *The Story of Dos Cabezas.* Tucson: Westernlore Press, 1995.

"Geology and Ore Deposits of the Courtland-Gleeson Region, Arizona." *Bulletin No. 123.* Tucson: University of Arizona, 1927.

Hurlbut, Cornelius S. Jr. *Dana's Manual of Mineralogy.* 16th ed. New York: John Wiley & Sons, Inc., Revised 1953.

International Correspondence Schools. *A Textbook on Metallurgy of Gold, Silver, Copper, Lead, and Zinc.* Scranton, PA: International Textbook Company, 1908.

Jones, Bob. *The Tucson Show: A Fifty-Year History.* Tucson: The Mineralogical Record, Inc., 2004.

"Mineral and Water Resources of Arizona." *Bulletin No. 180.* Tucson: University of Arizona, 1969. Sonnichsen, C.L. *Colonel Greene and the Copper Skyrocket.* Tucson: University of Arizona Press, 1974.

Ring, Bob, Ring, Al, and Tallia Pfrimmer Cahoon: *Ruby, Arizona – Mining, Mayhem, and Murder.* Tucson: U.S. Press & Graphics, 2005.

Schrader, Frank C. *Mineral Deposits of the Santa Rita and Patagonia Mountains Arizona.* Washington, DC: Government Printing Office, 1915.

Sinkankas, John . *Gemstones and Minerals: How and Where to Find Them.* Princeton, New Jersey: D. Van Nostrand Company. Inc, 1961.

Sloane, Howard N., and Lucille L. *A pictorial history of American mining; the adventure and drama of finding and extracting nature's wealth from the earth, from pre-Columbian times to the present.* New York: Crown Publishing, 1970.

Stewart, Charles A. "The Geology and Ore Deposits of the Silverbell Mining District, Arizona." *Bulletin No. 65.* Tucson: University of Arizona, 1912.

Young, Otis E. Jr. *Western Mining an Informal Account of Precious-Metals Prospecting, Placering, Lode Mining, and Milling on the American Frontier from Spanish Times to 1893.* Norman: University of Oklahoma Press, 1970.

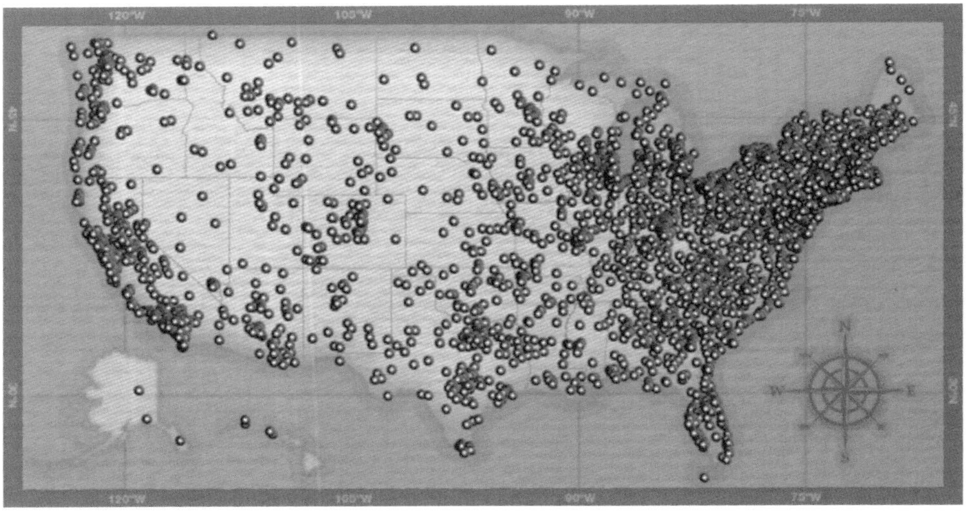